Proceedings

Ein stetig steigender Fundus an Informationen ist heute notwendig, um die immer komplexer werdende Technik heutiger Kraftfahrzeuge zu verstehen. Funktionen, Arbeitsweise, Komponenten und Systeme entwickeln sich rasant. In immer schnelleren Zyklen verbreitet sich aktuelles Wissen gerade aus Konferenzen, Tagungen und Symposien in die Fachwelt. Den raschen Zugriff auf diese Informationen bietet diese Reihe Proceedings, die sich zur Aufgabe gestellt hat, das zum Verständnis topaktueller Technik rund um das Automobil erforderliche spezielle Wissen in der Systematik aus Konferenzen und Tagungen zusammen zu stellen und als Buch in Springer.com wie auch elektronisch in SpringerLink und Springer Professional bereit zu stellen.

Die Reihe wendet sich an Fahrzeug- und Motoreningenieure sowie Studierende, die aktuelles Fachwissen im Zusammenhang mit Fragestellungen ihres Arbeitsfeldes suchen. Professoren und Dozenten an Universitäten und Hochschulen mit Schwerpunkt Kraftfahrzeug- und Motorentechnik finden hier die Zusammenstellung von Veranstaltungen, die sie selber nicht besuchen konnten. Gutachtern, Forschern und Entwicklungsingenieuren in der Automobil- und Zulieferindustrie sowie Dienstleistern können die Proceedings wertvolle Antworten auf topaktuelle Fragen geben.

Today, a steadily growing store of information is called for in order to understand the increasingly complex technologies used in modern automobiles. Functions, modes of operation, components and systems are rapidly evolving, while at the same time the latest expertise is disseminated directly from conferences, congresses and symposia to the professional world in ever-faster cycles. This series of proceedings offers rapid access to this information, gathering the specific knowledge needed to keep up with cutting-edge advances in automotive technologies, employing the same systematic approach used at conferences and congresses and presenting it in print (available at Springer.com) and electronic (at SpringerLink and Springer Professional) formats.

The series addresses the needs of automotive engineers, motor design engineers and students looking for the latest expertise in connection with key questions in their field, while professors and instructors working in the areas of automotive and motor design engineering will also find summaries of industry events they weren't able to attend. The proceedings also offer valuable answers to the topical questions that concern assessors, researchers and developmental engineers in the automotive and supplier industry, as well as service providers.

Weitere Bände in der Reihe http://www.springer.com/series/13360

Johannes Liebl
(Hrsg.)

Der Antrieb von morgen 2018

Der Wandel im Ökosystem –
prägend für den Antrieb
12. Internationale MTZ-Fachtagung
Zukunftsantriebe

 Springer Vieweg

Herausgeber
Johannes Liebl
Moosburg, Deutschland

ISSN 2198-7432 ISSN 2198-7440 (electronic)
Proceedings
ISBN 978-3-658-21418-0 ISBN 978-3-658-21419-7 (eBook)
https://doi.org/10.1007/978-3-658-21419-7

Die Deutsche Nationalbibliothek verzeichnet diese Publikation in der Deutschen National-
bibliografie; detaillierte bibliografische Daten sind im Internet über http://dnb.d-nb.de abrufbar.

Verantwortlich im Verlag: Markus Braun

Gedruckt auf säurefreiem und chlorfrei gebleichtem Papier

Springer Vieweg ist ein Imprint der eingetragenen Gesellschaft Springer Fachmedien Wiesbaden
GmbH und ist ein Teil von Springer Nature
Die Anschrift der Gesellschaft ist: Abraham-Lincoln-Str. 46, 65189 Wiesbaden, Germany

Vorwort

Durch die drängenden Herausforderungen des Klimaschutzes und der Luftreinhaltung befindet sich die Automobilindustrie in einem Paradigmenwechsel. Die Elektrifizierung schreitet weiter voran. Antriebsstränge müssen noch stärker im Systemverbund Verbrennungsmotor, Getriebe- und Elektrifizierung betrachtet werden. Das Duo aus Verbrennungs- und Elektromotor wird in der nächsten Dekade des Automobilbaus einen bisher nie dagewesenen Stellenwert einnehmen. Um hybride Antriebsstränge in unterschiedlichen Märkten und für verschiedene Fahrzeugsegmente darzustellen, ist es entscheidend, über die Komponentenebene hinaus das Gesamtsystem zu beherrschen.

Einer der Schwerpunkte der 12. Internationalen MTZ-Fachtagung *Der Antrieb von morgen* werden Energieträger, insbesondere optimierte Kraftstoffe, sein. Die Mobilitätswende ist ohne das intelligente Zusammenspiel von Elektrifizierung und synthetischen Kraftstoffen nicht zielgerichtet umsetzbar. Elektrische Komponenten, hier vor allem Speicher und Antriebe, bilden einen weiteren Schwerpunkt der kommenden Tagung. Die für die Elektrifizierung so wichtigen Speichermedien müssen robust ausgelegt sein, ohne die Kosten und den ökologischen Fußabdruck zu vernachlässigen. Die Spreizung der Hybridantriebe von 48-Volt- bis hin zur Hochvolttechnik eröffnet Möglichkeiten für alle Fahrzeugsegmente. Daher ist auch das elektrische Gesamtsystem eines der zentralen Themen. Hier schlagen wir den Bogen vom 48-V-Konzept bis hin zur Brennstoffzelle.

Im Namen aller Beteiligten und unsere Partner Schaeffler und Volkswagen lade ich Sie ein, bei der 12. Internationalen Fachtagung *Der Antrieb von morgen* dabei zu sein. Wir freuen uns schon heute auf den aktiven Dialog mit Ihnen!

Für den Wissenschaftlichen Beirat
Dr. Johannes Liebl
Herausgeber ATZ I MTZ I ATZelektronik

V

Editorial

The urgent challenges of protecting the climate and reducing air pollution are causing a paradigm shift in the automotive industry. Powertrains must be considered more than ever before as an integrated system consisting of an engine, a transmission and electrification. Over the next decade of automotive engineering, the combination of an internal combustion engine and an electric motor will gain unprecedented importance. In order to develop hybrid powertrains in different markets and for different vehicle segments, it is vital to master the overall system beyond the component level.

One of the focal points of the 12th International MTZ Conference on "The Powertrain of Tomorrow" will be energy sources, in particular optimized fuels. The mobility revolution cannot be successfully completed without intelligent interaction between electrification and synthetic fuels. Electric components, and above all batteries and motors, are another major topic at the conference. Batteries, which play such a key role in electrification, must be designed to be robust, but without neglecting their cost and ecological footprint. Hybrid drive systems span the range from 48-volt to high-voltage technology, thus opening up new possibilities for all vehicle segments. For that reason, the overall electric system is also one of the central topics. We cover all areas from 48-volt concepts to the fuel cell.

On behalf of all those involved as well as our partners Schaeffler and Volkswagen, I would like to invite you to attend the 12th International MTZ Conference on "The Powertrain of Tomorrow". We look forward to an interesting dialogue with you.

On behalf of the Scientific Advisory Board
Dr. Johannes Liebl
Editor-in-Charge ATZ I MTZ I ATZelektronik

Inhaltsverzeichnis

Autorenverzeichnis

Dr. Volker Ambrosius IAV GmbH, Radebeul, Deutschland

Thomas Arnold IAV GmbH, Mitteldorf, Deutschland

Dr. Martin Brüll Continental Automotive GmbH, Regensburg, Deutschland

Andreas Burkert Springer Fachmedien Wiesbaden GmbH, Wiesbaden, Deutschland

Siegmund Deinhard Continental Automotive GmbH, Regensburg, Deutschland

Alexander Ebel APL Automobil-Prüftechnik Landau GmbH, Landau, Deutschland

Dr. Alfred Elsäßer MAHLE International GmbH, Stuttgart, Deutschland

Friedrich Graf Continental Automotive GmbH, Regensburg, Deutschland

Peter Haußmann Ruhr-Universität Bochum, Bochum, Deutschland

Fuliang Huang Keihin Corporation, Tochigi, Japan

Akira Ichinose Keihin Corporation, Tochigi, Japan

Andreas Kemle MAHLE International GmbH, Stuttgart, Deutschland

Prof. Dr. Manfred Klell HyCentA Research GmbH, Graz, Österreich

Matthias Krause IAV GmbH, Mitteldorf, Deutschland

Vincent Lawlor AVL List GmbH, Graz, Österreich

Christian Lensch-Franzen APL Automobil-Prüftechnik Landau GmbH, Landau, Deutschland

Prof. Dr. Joachim Melbert Ruhr-Universität Bochum, Bochum, Deutschland

Tobias Mink APL Automobil-Prüftechnik Landau GmbH, Landau, Deutschland

Masashi Murohoshi Keihin Corporation, Tochigi, Japan

Dr. Heiko Neukirchner IAV GmbH, Mitteldorf, Deutschland

Thomas Pfund LuK GmbH & Co. KG, Bühl, Deutschland

Jürgen Rechberger AVL List GmbH, Graz, Österreich

Michael Reissig AVL List GmbH, Graz, Österreich

Daniel Rieger MAHLE International GmbH, Stuttgart, Deutschland

Martin Schäfer APL Automobil-Prüftechnik Landau GmbH, Landau, Deutschland

Dr. Otmar Scharrer MAHLE International GmbH, Stuttgart, Deutschland

Tingting Sui Keihin Corporation, Tochigi, Japan

Dr. Alexander Trattner HyCentA Research GmbH, Graz, Österreich

Dr. Marco Warth MAHLE International GmbH, Stuttgart, Deutschland

Electric Axle Drives – scalable propulsion system for electrified powertrains

Dipl.-Ing. **Thomas Pfund**, LuK GmbH & Co. KG, Buehl, Germany

© Springer Fachmedien Wiesbaden GmbH, ein Teil von Springer Nature 2018
J. Liebl (Hrsg.), *Der Antrieb von morgen 2018*, Proceedings,
https://doi.org/10.1007/978-3-658-21419-7_1

Zusammenfassung

Bei der Elektrifizierung des Antriebsstranges spielen elektrische Achsen eine entscheidende Rolle. Der Einsatz ist sowohl für hybridische als auch für rein elektrische Antriebskonzepte sinnvoll, wodurch sich anwendungsbezogen sehr verschiedene Leistungs- und Bauraumanforderungen ergeben. Funktional sind verschiedene Optionen, wie z.B. Parksperre, Differentialsperre oder Torque Vectoring zu berücksichtigen. Zur Optimierung von Bauraum, Gewicht und Kosten zeigt sich ein klarer Trend, Getriebe, Elektromotor und Leistungselektronik zu einem Aggregat zu integrieren. Dazu sind diese Hauptkomponenten so zu modularisieren, dass eine einfache Skalierbarkeit bei gleichzeitiger Optimierung des spezifischen Bauraums möglich wird.

Auf der Basis des Schaeffler Leichtbaudifferentials lassen sich Getriebelösungen mit hoher Leistungsdichte in koaxialer oder achsparalleler Ausprägung darstellen. Für den elektrischen Antrieb wurde ebenfalls ein skalierbares Modulkonzept entwickelt. Neben Magnetkreisen unterschiedlicher Technologie wurden für die Leistungselektronik Baugruppen definiert, die die hohen Anforderungen an Zyklenfestigkeit, Leistungsdichte und Robustheit in Bezug auf die Umgebungsbedingungen im Aggregat erfüllen. Der Beitrag zeigt den Aufbau der mechanischen und elektrischen Module und deren funktionale und physikalische Integration in eine spezifische elektrische Achse.

Abstract

In electrification of the powertrain, electric axles play a decisive role. They are appropriate for use both in hybrid and in pure electric drive concepts which, as a function of the application, have very different performance and design envelope requirements. In functional terms, various options such as parking lock, differential lock or torque vectoring must be taken into consideration. For optimization of the design envelope, weight and costs, there is a clear trend towards the integration of the transmission, electric motor and electronic power system in one assembly. To this end, these principal components must be modularized in such a way that scalability is easily achieved while optimizing the specific design envelope.

On the basis of the Schaeffler lightweight differential, transmission solutions of high performance density can be presented in a coaxial or axially parallel design. A scalable modular concept was also developed for the electric drive. In addition to magnet circuits employing various technology, power electronics modules were defined that fulfill the high requirements for cycle strength, performance density and robustness in relation to the ambient conditions in the assembly. The paper shows the design of mechanical and electric modules as well as their functional and physical integration in a specific electric axle.

1 Review

With the electrically powered prototype vehicle Active eDrive, the Schaeffler Group presented a concept vehicle in 2011 containing a powertrain that then pointed the direction to the future. At Schaeffler, a Skoda Octavia Scout was converted into an electrically driven four-wheel drive vehicle that, in addition to the electric drive, facilitates intelligent electromechanical torque distribution over both the front and rear axle. The inverters for the traction and torque vectoring drives and the powertrain control unit were shown as prototypes in this demonstrator.

An essential USP of the Schaeffler e-axle was its coaxial design. The electric motor used was a PSM machine from IDAM (INA Drives & Mechatronics) designed as a hollow shaft motor with a maximum speed of 14 000 rpm.

The transmission was already based on the Schaeffler lightweight differential of a planetary design. The differential was integrated such that the output was delivered via the planetary carrier and one of the sun gears.

This gave the possibility of linking the output shafts to a superimposing gear set that is, on the input side, connected to a servo motor.

The differential, the superimposed transmission and the servo motor only rotate while cornering and stay stationary during straight line travel. As a result, losses are minimized. In deactivated state of the servo motor the differential acts like a standard differential with a somewhat higher lock-up value. The power rating of the servo motor was 7kW, allowing a differential torque of 1200Nm.

This vehicle was very important for Schaeffler, not only in the development of axle technologies but also in the analysis of their behavior in the vehicle (acoustics, control strategies, package etc.).

2 Market, powertrain concepts and resulting requirements

Electric axles have now become established as a drive system and are used in powertrains for hybrid and electric vehicles. There are numerous possibilities for integrating an e-axle in functional terms. In hybrid powertrains, the e-axle can potentially be assigned the following principal functions:

- optimization of consumption (CO_2) through utilization of kinetic energy (regeneration)
- electric driving in a limited speed range (in the case PHEVs, up to 120km/h)
- electric all-wheel drive, optionally with torque vectoring

Since the electric drive power in the hybrid powertrain is generally less than that from a combustion engine, the combination of these functions leads to a conflict in objectives. Regeneration and all-wheel drive should be available over the whole speed range. In addition, the vehicle should nevertheless offer dynamic electric travel. A solution to this conflict, which is intensified significantly in the case of heavy vehicles, is the use of a twin stage transmission. In this case, the first stage is designed such that purely electric travel can be covered up to approx. 120 km/h. Such a solution is particularly suitable for hybridization in the strongly growing SUV segment (Figure 1).

Shifting into second gear is only performed when the combustion engine is running, so that the transmission does not need to be dependent on power-shift. It is possible to use an economical synchronized gearshift such as is well known in the case of manual transmissions.

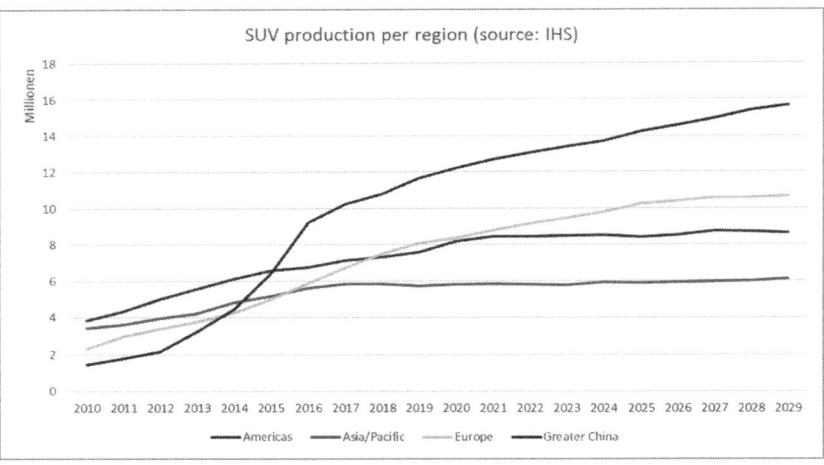

Fig. 1: Regional SUV market growth

In the case of powertrain architectures where a second electric machine is integrated on the axle driven by the combustion engine in such a way that it can generate an additional traction torque (DHT, P2), a single stage transmission is adequate. An e-axle of this type can also be used in pure electric vehicles as a secondary or all-wheel conversion drive. Power lies in the range between 50kW and 100kW. Asynchronous machines (ASM) contribute to the minimization of drag losses. A somewhat larger design envelope is of course necessary. In addition, rotor cooling is usually required.

Higher power levels must be provided for the main drive in electric vehicles. Depending on the vehicle mass and market segment, power levels between 120kW and 300kW are required. Single stage transmissions can be used for maximum speeds up to 200km/h. For higher speeds, twin stage transmissions should also be considered. In this case, however, interruption of traction power during gearshift cannot be tolerated.

Due to the wide application spectrum of electric axles and the increasing electrification of powertrains, significant growth rates can be anticipated. The precise rate at which hybrid and electric powertrains will come into use is difficult to forecast since there is an interaction of many influences like legislation on consumption and emissions, regional travel restrictions, developments on battery prices, alternative fuels, regenerative energies, infrastructure, customer acceptance etc. Schaeffler is therefore working on alternative scenarios. Figure 2 shows the increase in the baseline scenario and the accelerated scenario.

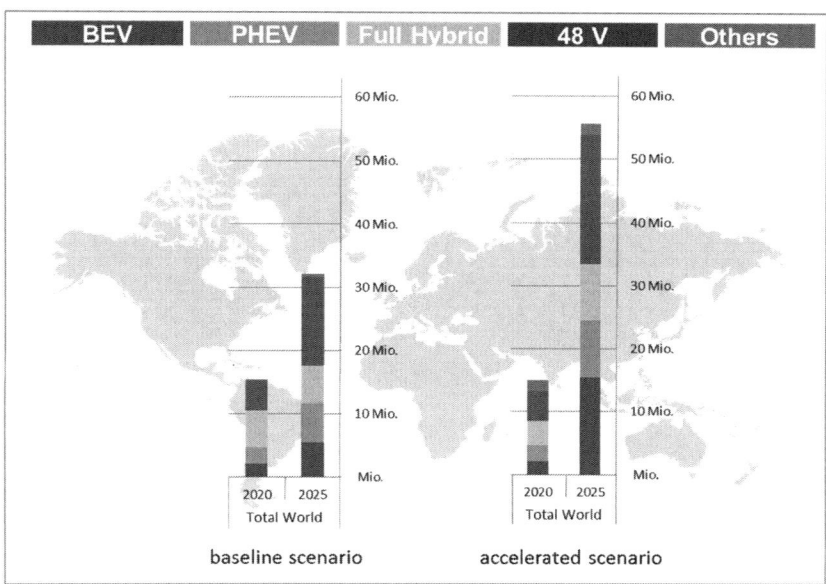

Fig 2: Alternative market growth scenarios

In addition to the principal travel functions, further requirements will come into play across applications. These include, on the one hand, optional functions such as a parking brake or torque vectoring. On the other hand, particular attention must be paid to acoustics since, during electric travel, operating noises are no longer masked by the combustion engine. Other factors determining the concept are functional safety, the cooling concept (optimum heat removal with minimal pressure drop), energy consumption and of course weight, design envelope versatility and to a very large extent the costs.

3 System Development

In order to develop system functions that are robust over the service life, load profiles for systems and components as well as acoustic or cooling behavior, analysis at the vehicle level is necessary. Schaeffler has lined up for the development of drive units on all levels of requirements and integration. In this way, numerous issues concerning longitudinal, transverse and vertical dynamics can be modeled in simulation.

The model levels are broken down into components in accordance with the system architecture and used with increasing depth of model accuracy for design and verification.

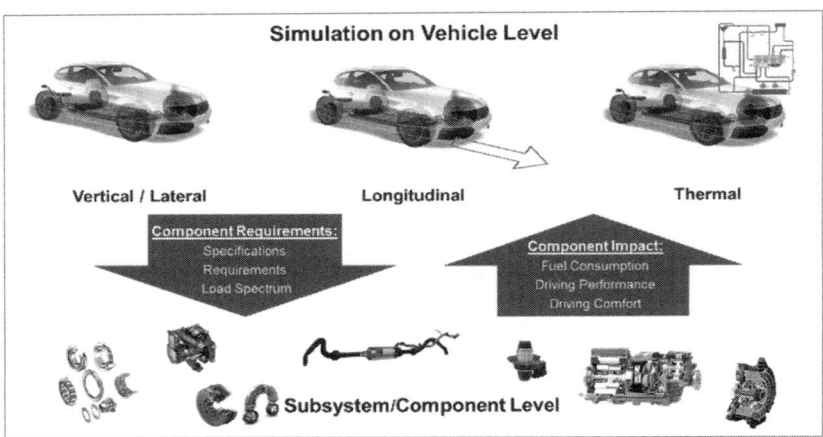

Fig 3: Simulation on vehicle level

An example of a classic system issue can be represented in this respect by torque accuracy. In the vehicle, it is essential that the required wheel torque can be provided with sufficient accuracy. Since the torque is not measured directly, a very large number of influences must be analyzed and assessed. As far as possible, these influences are taken into consideration in the control model and the parameters are either adapted online or are determined by metrological methods on the EOL test rig. Figure 4 shows, as an example, the result of a sensitivity analysis on a specific drive system.

The graph shows air gap torque as a function of phase current (dotted line) with deviations (point cloud) caused by other influencing effects (i.e. tolerances, measurement accuracy).

Schaeffler has selected the architecture and definitions of subsystems in such a way that flexibility in relation to the package, system performance and functions is maintained. In the combination of optimization criteria (acoustics, mechanical strength, oiling etc.), the focus is always on designing the electric machine with the highest possible speed and using the resulting benefits in terms of design envelope, weight and costs.

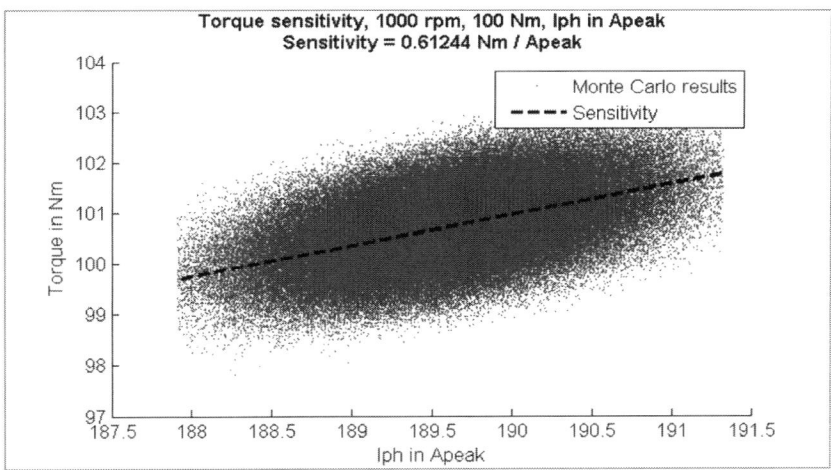

Fig. 4: Sensitivity Analysis – torque accuracy

4   Mechanical Transmission

In order to optimize the energy conversion chain and match it to the specific vehicle, the ratio of the transmission must be adaptable in a range between i=8 and i=18. At the same time, it must be possible to respond to the application-specific package and robust solutions must be implemented in relation to acoustics, oiling and heat removal. For this purpose, Schaeffler has defined base gear sets that can be matched to the specific requirements. A basic distinction is made between two speed transmissions for hybrid applications and single speed transmissions for hybrid and electric vehicles.

In order to present an optimum package, the two speed transmission comprises a planetary set on the electric motor side that is deactivated in the second stage by means of synchronized gearshift. For the gearshift actuator, Schaeffler has drawn on its in-house portfolio for automation of transmissions. If higher ratios are required, an intermediate shaft can be added.

Fig. 5: Two speed e-axle for hybrid applications

For the strongly growing market in electric axles for hybrid and electric vehicles, the single speed gear set can be of an extremely compact construction due to the lightweight differential. In this application, the classic differential bevel gears are replaced by at least six differential spur gears and the output bevel gears by two sun gears. The differential gears are arranged in pairs on the circumference of the sun gears. One differential spur gear in each pair is linked with the left hand sun gear, while the other gear is in tooth contact with the right hand sun gear. An essential characteristic of the "Schaeffler lightweight differential" is the arrangement of the tooth contacts in two planes, which is realized by means of a different profile shift on the sun gears. As a result, there is a ratio of -1 between the output shafts where the sun gears have the same number of teeth.

The lightweight differential is advantageous particularly in the coaxial design of the drive system, since it has a very short axial length and can be directly nested with the input stage of the transmission. The differential housing can be used simultaneously as planetary carrier of the input stage. The differential does not therefore require its own drive wheel but is driven directly via the planetary carrier. In the complete transmission, this gives a minimum ratio that corresponds to the stationary ratio plus 1. If a further planetary gear stage is integrated, the overall ratio depending on the coupling method of the planetary gear stages to each other increases by a maximum factor "standard ratio plus 1".

A further increase in performance density is achieved if a stepped planet set is arranged ahead of the differential. Figure 6 shows a gear set that combines a planetary stage interleaved with the spur gear differential. This highly compact design of a complete gear set offers a transmission ratio of approx. i=8.

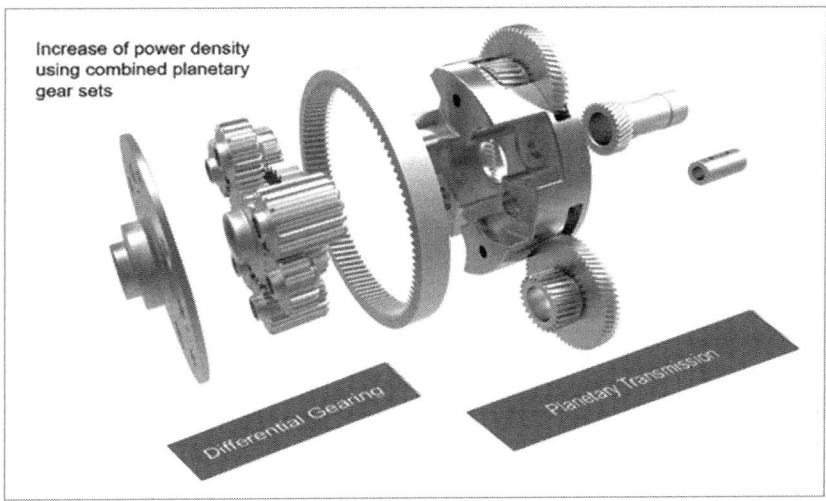

Fig. 6: Stepped planetary gear set interleaved with spur gear differential

If higher transmission ratios are required, a second planet set is arranged ahead, for example in a power-split arrangement.

Building on these relationships, Schaeffler has defined a portfolio of gear sets that is always based on the same base elements but nevertheless offers a wide range of transmission ratios. Figure 7 shows a schematic of the different transmission variants of co-axial design, building on each other. The transmission scheme shows only the planetary gear stages in the transmission without the differential. The relationship between the ratio of the transmission and the size of the electric motor is shown here in a highly simplified form.

Existing package studies at Schaeffler have shown that the coaxial design of the e-axle is advantageous in many cases for remaining within the specified design envelope limits and at the same time presenting scalable solutions for various performance classes within one platform.

For those cases in which an axially parallel concept is necessary, the differential can be coupled by means of an additional spur gear stage.

Furthermore, the lightweight differential in the layout type described above offers the possibility of coupling a servo motor via an additional superimposing gear set. As a result, the option of "torque vectoring" can be offered.

Solutions are also available for the integration of a parking brake.

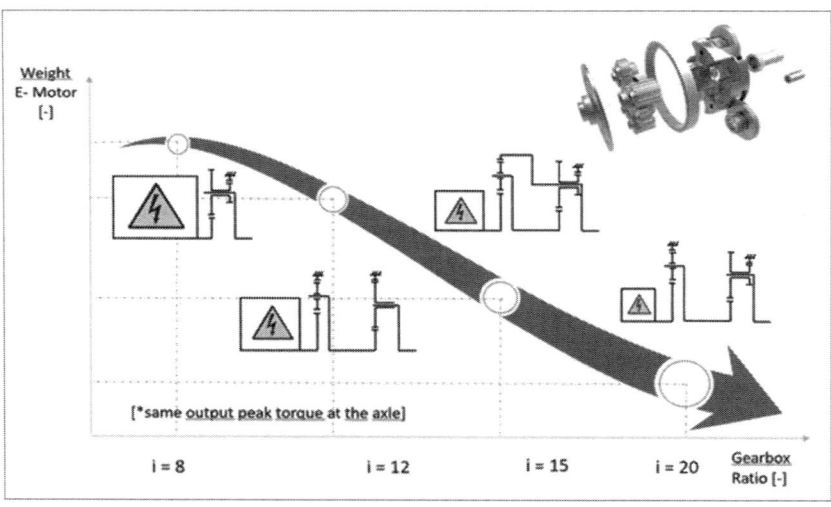

Fig. 7: Basic gear sets for co-axial applications

5 Electric Motors

Due to the clear trend towards highly integrated assemblies, the electric motor should no longer be understood as an independently functioning unit. Instead, scalable active parts are required for the stator and rotor, whose design should be seen as an optimization problem in terms of different and in some cases opposing criteria. The essential optimization criteria are:

– performance density
– loss reduction/efficiency
– cooling
– minimization of harmonic excitations due to torque fluctuations or magnetically induced radial forces
– costs

If high performance densities are required, permanent magnet synchronous machines (PSM) are used.

For reasons of optimum material utilization (winding factor, use of reluctance torque), the minimization of rotor losses and the reduction of acoustic excitation by harmonics, the use of distributed winding is advisable. The rapid development of manufacturing technology has now made it possible to produce "constructed" HV windings in high volumes. The most well-known variant is hairpin winding. In addition to the advantages stated at the beginning, the copper fill factor is significantly increased and the winding head is shortened in such a way that the losses in the magnetically inactive part of the winding no longer cause significant impact (under the precondition of sufficient active length).

Particular attention must be paid, however, to the current displacement effects, which particularly affect the bars near the air gap. This phenomenon must be taken into consideration by targeted measures in the design, since the resulting losses cannot be disregarded.

In this machine design (PSM), rotor cooling can usually be omitted, since the main losses occur in the stator. In most cases, stator cooling is carried out by means of a cooling duct on the outermost circumference of the stator using a water/glycol mixture. The design must ensure that the outer jacket undergoes uniform flow at all temperatures (prevention of hotspots) and that the pressure loss remains at a minimum. At Schaeffler, CFD simulation is an established component of the design chain.

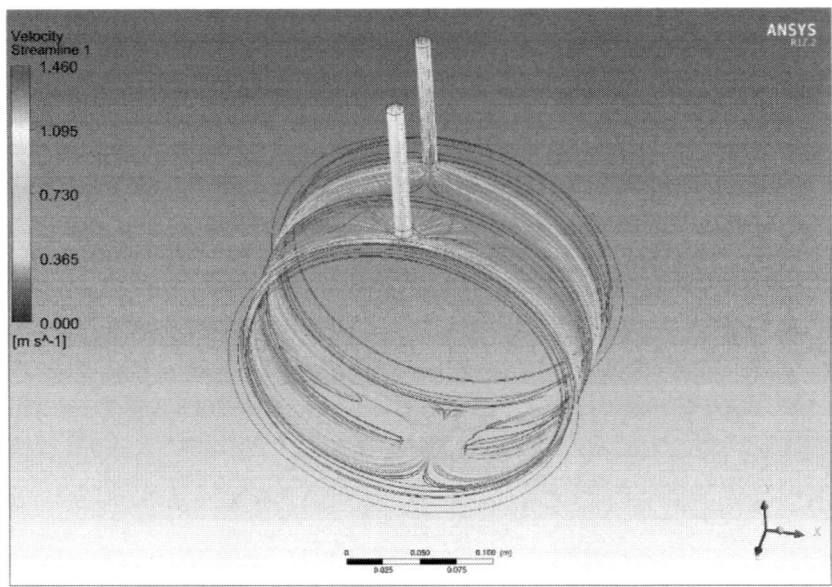

Fig. 8: Cooling channel – CFD flow optimization

In some cases, the use of an asynchronous machine is required. The rotor in an ASM is, to begin with, more economical to manufacture than that of a PSM. It must be noted, however, that rotor cooling is generally necessary in the case of the ASM. It is possible to use the transmission oil for this purpose. Additional outlay, for example on an oil pump and cooler, reduces the cost advantage of the ASM.

In order to cover the requirements arising from the various e-axle applications, basic modules for the stator and rotor were defined at Schaeffler. The stator has bar winding and can be scaled in two diameter classes by means of length and variation of the winding. This allows matching to the specific design envelope. The matching of performance is achieved by a combination of various options from the modular technology concept (rotor topology as PSM or ASM, oil cooling, rotor cooling, slot cooling). Figure 9 shows this approach.

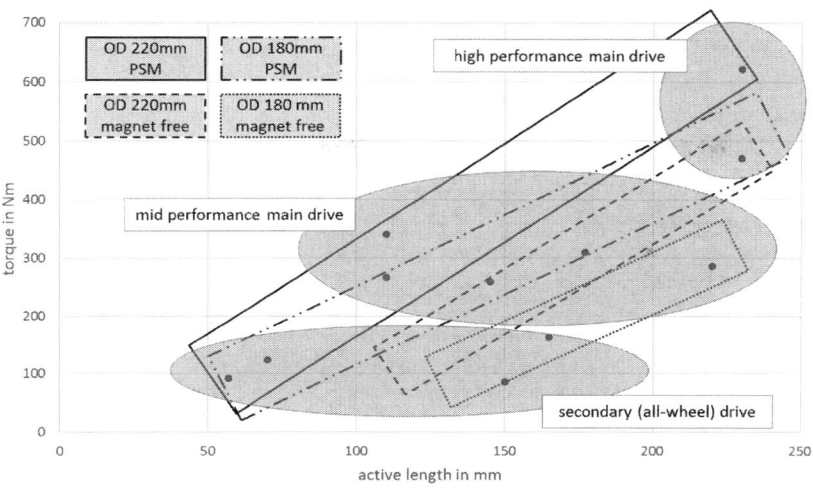

Fig. 9: Power range of electric machines

6 Power Electronics

The inverter must also be integrated in the axle assembly. Platform solutions are now planned for various vehicle classes and large quantities. This means that the load and ambient conditions must be specified with greater variance. This leads to particular requirements in relation to design envelope versatility, robustness and costs. Failure mechanisms, such as those in bonding wires or solder connections must be precisely analyzed and, where possible, eliminated by means of alternative solutions. The architecture must be selected such the system is functionally secure (ASIL C or D) and the software must allow the integration of customer-specific modules.

For these reasons, Schaeffler has also defined the inverter as a modular concept. The principal components here are:

- control board with Aurix TC277 and watchdog as well as control software based on an Autosar 4.x stack
- power board with gate drivers and various additional functions
- power module with carrier frame, cooling structures, lead frames etc.
- half bridge modules

These components can be arranged and integrated within the scope of defined rules. In relation to functional safety, a three level concept is implemented. In order to achieve maximum robustness and the conditions for optimum heat removal, sintered half bridge modules are used.

Scalability can be managed in two ways. One possibility is the silicon population of the individual half bridge module. The objective here is to have as few variants as possible in order to minimize costs and risks in development. A further possibility lies in the use of several modules as a parallel circuit or for multi-phase designs of the drive.

In addition to the circuit, functional and EMC-optimized integration, the cooling concept is an essential factor. As in the case of the electric motor, cooling is carried out as standard using a water/glycol mixture. At minimal pressure loss, cooling over the specified temperature range must be ensured. Optimization is also carried out here by means of a CFD simulation (Figure 10). The quality of simulation has now become so high that absolutely accurate statements can be made.

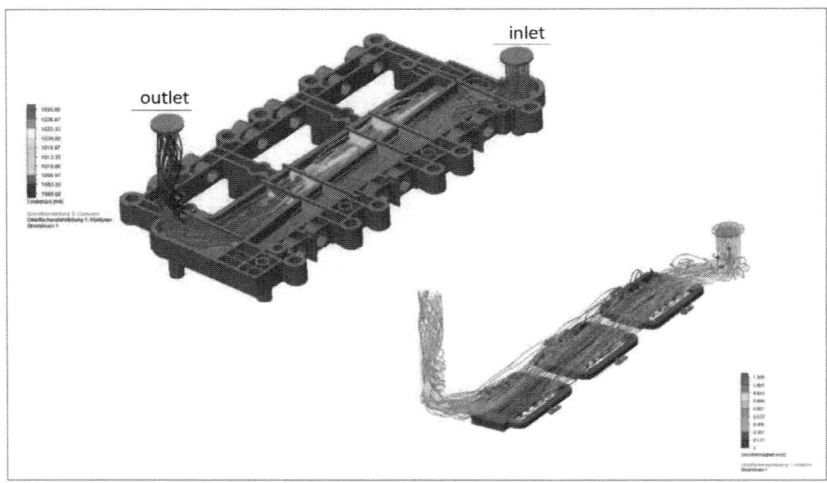

Fig. 10: Carrier frame – CFD flow optimization

Figure 11 shows an example from the modular concept described as a 550A variant. In this case, the control and driver part was implemented on one circuit board.

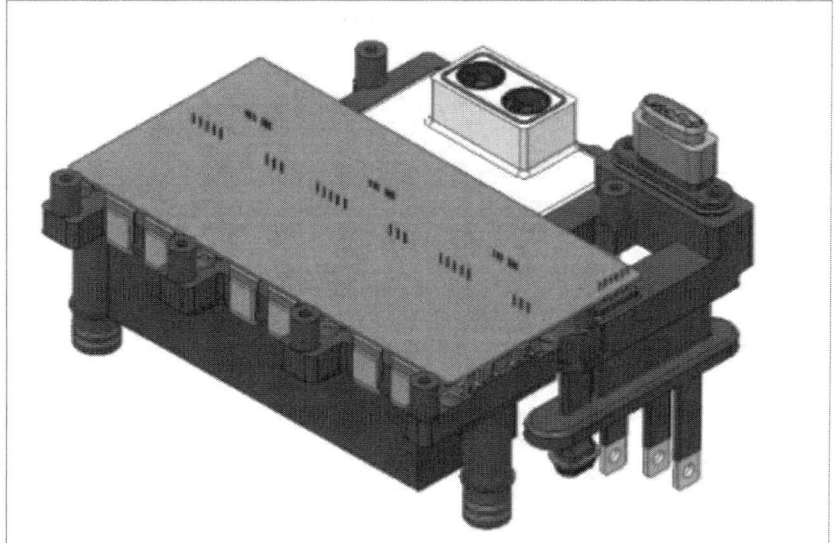

Fig. 11: 550A inverter

7 Summary

Starting from the first functional prototype "Active eDrive" in 2011, Schaeffler has worked systematically on the optimization of the axle on all system levels. In addition to technical further development, the development processes and the necessary tool chain have been developed and implemented. The competences that were initially lacking, particularly in the field of the electric drive and the associated software, have been added. The final figure 12 gives an impressive vision of what is possible with the state of the art.

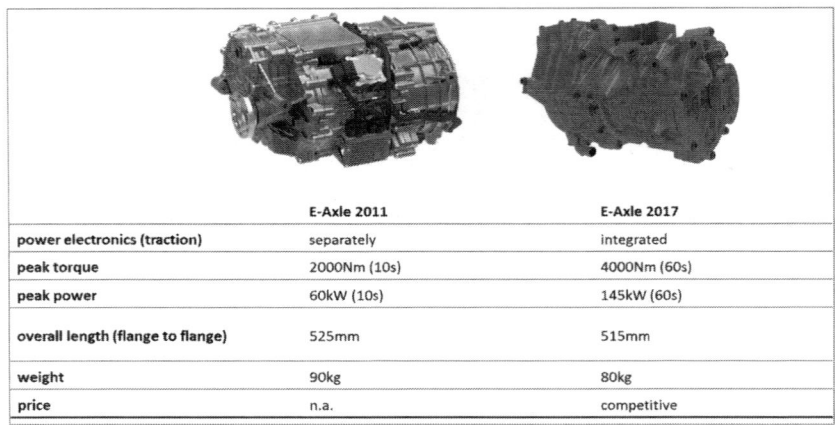

	E-Axle 2011	E-Axle 2017
power electronics (traction)	separately	integrated
peak torque	2000Nm (10s)	4000Nm (60s)
peak power	60kW (10s)	145kW (60s)
overall length (flange to flange)	525mm	515mm
weight	90kg	80kg
price	n.a.	competitive

Fig. 12: Schaeffler e-axle 2011 and 2017

8 Literaturangaben

[1] T. Smetana, M. Berger, M. Gramann, M. Mitariu-Faller: Modular System for electrical drive axles. ATZ 09/2013, Volume 115, S. 60-65

[2] T. Pfund, M. Gramann, M. Fritz, E. Enderle: Integrierte Leistungselektronik eröffnet Effizienzpotenziale. ATZ elektronik 05/2013, 8. Jahrgang, S. 360-367

MAHLE Efficient Electric Transport – an efficient system solution for the electrified urban mobility

Daniel Rieger

Andreas Kemle

Dr. Alfred Elsäßer

Dr. Marco Warth

Dr. Otmar Scharrer

© Springer Fachmedien Wiesbaden GmbH, ein Teil von Springer Nature 2018
J. Liebl (Hrsg.), *Der Antrieb von morgen 2018*, Proceedings,
https://doi.org/10.1007/978-3-658-21419-7_2

1 Introduction

Driven by numerous trends and challenges the automotive industry faces a tough transformation. This affects car manufactures as well as the whole supplier industry. The main challenge thereby is the further reduction of CO_2 emissions. During the UN climate conference 2015 in Paris 197 nations agreed to collaborate on common climate change politics. One outcome of this is to empower the climate change by limiting the global temperature rise the global temperature rise compared to the pre-industrial level to 2 °C within this century [1]. This target can only be reached by the collaboration of all branches of industries to reduce CO_2 emissions. The automotive industry (incl. commercial vehicles and passenger cars) has a share of roughly 10 % of the overall CO_2 emissions [2]. The current limit for CO_2 emissions for passenger cars in Europe is 130 g/km from the beginning of 2015 and will be reduced to 95 g/km in 2021 [3].

Beside the CO_2 challenge another big task is the air pollution within large metropolitan areas worldwide. Here the critical emission values are frequently exceeded. As a consequence, cities like Paris are planning to ban vehicles with combustion engines from 2030 onwards [4]. In London it will get more expensive to drive with them into the city [5]. And even entire countries like Slovenia or the Netherlands intend to prohibit these vehicles [6], [7].

Both challenges for the automotive industry lead to the need for new mobility concepts and powertrain technologies. Another trend on top of the CO_2 and the local air pollution discussion within large metropolitan areas adds a key aspect to MAHLE's new concept vehicle: advancing urbanization. By the year 2050 roughly 70 % of the world population will live in large cities [8]. This trend even increased the ongoing challenges for the automotive industry.

Given these challenges, MAHLE developed a "new-thinking", fully electric urban vehicle concept called MEET – MAHLE Efficient Electric Transport. Combining its core competence and expertise both in electric powertrains and thermal management, as well as climate comfort, MAHLE developed an efficient system solution for the electrified urban mobility.

2 Urban Mobility at MAHLE

In order to meet the governing trends and challenges for the automotive industry, as well as to create a costumer benefit in urban mobility, an entirely new vehicle concept is required. MAHLE defined key aspects of its urban mobility concept, right at the start of the development (see Figure 1 on the left side).

By reason of legislation but as well as by personal belief sustainability and environmental awareness are recognized by the population. One major issue to cover these topics is the choice of the appropriate powertrain. In addition a suitable application for urban traffic in terms of power requirement, comfort and costs is necessary to deal with the subject efficiency holistically. Digital technologies are important in the context of urban mobility increasing comfort and enabling intermodality. Comfort for example could mean how to control your vehicle and the integration of a smartphone within the vehicle. The user gets both the latest information on the vehicle condition and digital technologies provide up-to-date intermodal mobility proposals with a universal access as well as payment systems for all ways of transport. Future mobility concepts are more than driving from A to B with one vehicle, but merely an intermodal and interconnected transportation system highlighting the importance of digital technologies as a key factor for urban mobility [9].

Figure 1: MAHLE's perspective on urban mobility

Figure 1 also shows the systematic approach for the development of the MEET vehicle concept. MAHLE derived the main consumer benefits and characteristics for its urban vehicle concept from the key factors identified. MEET is a battery electric vehicle. It is a system approach for a sustainable mobility free of local emissions and CO_2. From a technology point of view the vehicle concept is developed to reach highest energy efficiencies, using a holistic approach combining the electric powertrain, thermal management and user-friendly intuitive operation. The interaction of numerous MAHLE technologies offers the potential to maximize efficiency and thus increasing the electrical driving range as one of the main consumer requirements using a smaller battery capacity as well as offering low cost and low weight.

3 MEET – a new experience of urban mobility

The urban vehicle concept MEET is an outlook from MAHLE how future mobility might look like and it shows a holistic approach for an efficient system solution for an electrified urban mobility. The entire vehicle concept was developed with the target of high system efficiency in mind. Efficiency in context of an electric vehicle is directly linked to the driving range. Thus a smaller and lower-cost battery can be implemented. MEET shows the possibility of a vehicle concept with 7 days driving without charging, owing to an overall range of 194 km in a typical urban drive cycle and the assumption of an average commuting distance of 20 kilometers per work day. Furthermore a MEET driver could also drive additional miles at the weekend with the same battery charge.

Climate comfort plays a huge part when striving for high system efficiency. It enables the MEET driver to use his car without major cutbacks in the electrical driving range at cold and hot ambient temperatures. Furthermore, the climate comfort functions within the MEET are 100 % individually adjustable as well as environmentally friendly.

The high efficiency electrical powertrain of the MEET is called 48 Volt Twin Power. It consists of two electric motors arranged on the rear axle of the vehicle. Both motors deliver 30 kW peak for transient driving situations. Fun to drive is delivered thanks to a sound agility with good acceleration, urban-dedicated maximum velocity as well as high maneuverability.

MEET is also affordable. The 48 Volt approach leads to a cost benefit of 25 % compared to a high-voltage powertrain with equal 30 kW power output [10]. Reduced high voltage safety features, a smaller battery management system and the electronics are main drivers for the lower cost. Thanks to a voltage level below 60 Volt, the whole concept is more user-friendly in maintenance for the driver or owner as well.

To achieve the four drivers or owner benefits MAHLE implemented the key technologies of the holistic MEET vehicle concept as shown in Figure 2.

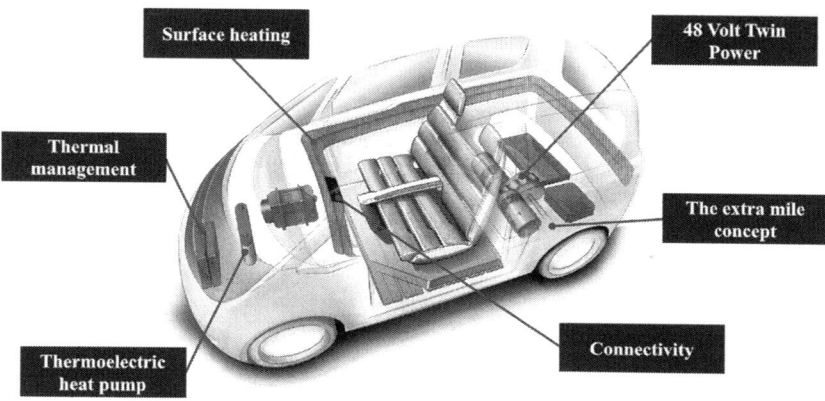

Figure 2: Main technologies of MAHLE's urban vehicle concept MEET

High system efficiency is inherently achieved by the holistic systems design and application throughout the whole vehicle concept. Both 48 Volt electric motors are operated with highest efficiency when running at urban speeds and loads. In addition the integrated power electronic, short distances to the battery and the thermal management of the powertrain reduce the overall losses. Thus the thermal management of the powertrain helps to increase the power density.

Another key contribution to the high system efficiency is the interplay of the several MAHLE technologies. One example is the use of any waste heat from the powertrain to efficiently heat up the cabin with a thermoelectric heat pump. Surface heating in the interior further adds to this innovative way to achieve climate comfort. The reduction of electrically heated air blowing into the cabin thanks to the heated surfaces enables further potential in increasing the overall efficiency and thus comfort of passengers.

Climate comfort plays a major role and another example how MAHLE addressed this in the MEET concept is via connectivity technologies. The urban vehicle concept can be controlled using a newly developed and intuitive HMI (human-machine-interface). The seamless integration of a smartphone further offers the possibility to personalize the comfort settings even before a driver enters the car. In the boot of the MEET concept car MAHLE shows an advanced solution for the extra mile from the parking space to the destination. Two Monowheels, charged via the vehicle battery enable people to travel comfortably and electrical even in pedestrian zones.

4 MEET Technologies

MEET offers a wide spread of innovative developments for a vehicle concept to address key factors in urban mobility. The main technologies are the electrical powertrain and the thermal management. Together they provide both high system efficiency and low cost structure as enabler for the electric mobility. The technologies described are novel developments for a specific design and application in urban traffic.

4.1 48 Volt Twin Power

MAHLE developed with the 48 Volt Twin Power a dedicated powertrain for the urban vehicle concept MEET. It is a fully electric powertrain to meet the key factors in urban mobility: sustainability and driving pleasure given the high efficient electric propulsion and affordability thanks to the low voltage level.

MAHLE derived the 'real world' requirements for such a powertrain from a characteristic drive cycle in Stuttgart. It consists of a challenging topology with varying traffic volumes. Extensive test drives on this distinctive city / urban mobility route (see Figure 3 left side) generated a comprehensive set of data to translate 'real world' driver demands to an electric powertrain specification. More details on the MAHLE Stuttgart Cycle can be found in [10]. This data was the input for a complete vehicle simulation using the specific MEET vehicle model. The power requirement within the Stuttgart Cycle is represented in Figure 3 on the right hand side. The plot shows the time-based frequency distribution of the motor torque and vehicle speed required.

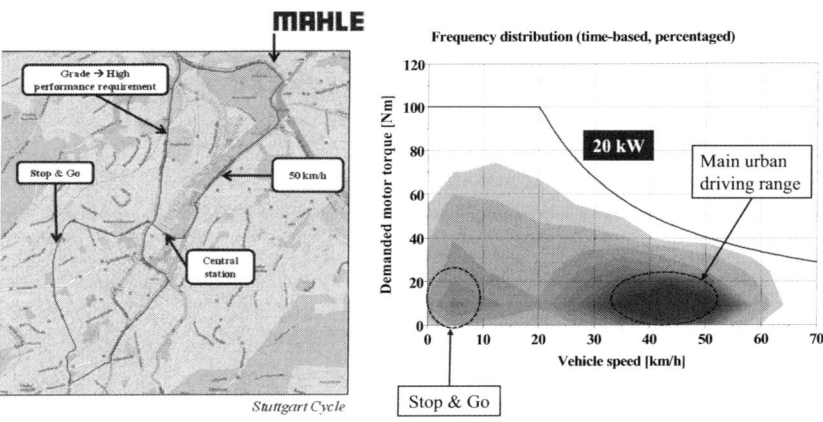

Figure 3: Stuttgart Cycle – Overview and Analysis

This analysis shows several general requirements for the design of the 48 Volt Twin Power propulsion system. There are two main driving operating conditions within the urban drive cycle. First there is the start / stop and sailing conditions at low vehicle speeds. The second important area is the velocity section between 40 and 50 km/h as the main operating range within the Stuttgart Cycle. For both driving conditions 20 kW of continuous mechanical power would be sufficient and 100 Nm of motor torque combined with an appropriate gearbox ratio also covers all city driving operating conditions.

The correct battery supply voltage to deliver the required output power is one of the main drivers for an affordable, i.e. low cost vehicle concept. MAHLE investigated the cost of electric drive systems for different power levels and supply voltages. Low voltage solutions with a voltage level of less than 60 Volts were compared with high voltage drives for the same power levels. With the results described in [10], a supply voltage of 48 Volt is chosen as the most suitable solution for the urban vehicle concept.

A 20 kW mechanical continuous 48 Volt solution offers a 25 % cost benefit of the propulsion system compared to an equally as powerful high voltage propulsion system. The 48 Volt low voltage system can be realized without the need for safety features like insulation observer or service disconnect for maintenance. The current power supply developments due to the increased electrical output demands for future hybrid vehicles changing from 12 Volts to 48 Volts further supports this decision. This change will enable using identical components with the advantage of economies of scale.

The 48 Volt Twin Power propulsion system is shown in Figure 4. It consists of two electric drive units with a central transmission. Each unit comprises both an electric motor and a power electronic, which are both liquid-cooled. This setup offers a great flexibility in positioning given the modular assembly and can be applied to various vehicle concepts. It does not require a mechanical differential, as each electric drive unit directly drives a wheel. Newly developed software functions electronically control different wheel speeds and torques.

Figure 4: MAHLE 48 Volt Twin Power

The electric motors are permanent magnet synchronous machines (PMSM) with a nominal supply voltage of 48 V and a maximum power output of 30 kW. The maximum power thereby can be delivered up to one minute. The PMSM has a special rotor design to use significant amounts of reluctance torque to deliver high torque values at low motor speeds. This is also required for the high power output at low supply voltage levels. The maximum speed of the 48 Volt Twin Power is 12.000 rpm for urban mobility applications. Attached to each electric motor is the power electronic. It consists of a connection terminal for the motor phases and DC voltage as a first layer. The current is guided through the logic board (second layer) which contains the current measurement, the microcontroller and the electric protection functions. The third layer is a heat exchanger for cooling the power and logic boards. The power board is the fourth and last layer of the electronics and includes the semiconductors. Due to the low voltage level the MOSFET technology can be used for the semiconductors, rather than the IGBT technology. This offers advantages in reduced switching losses [11].

The 48 Volt Twin Power motors in combination with the integrated power electronics enable both a high system peak efficiency of > 93 % and efficiencies in excess of 90 % for the entire main urban driving range.

High agility in urban traffic like small turning circle and an easy to handle parking in narrow situations are important characteristics for an urban vehicle concept. These benefit in the presented powertrain configuration from an all-electronic torque-vectoring functionality. The propelled wheels can be specifically controlled with different speed and torque settings. The central transmission shown in Figure 4 has a single gear ratio of 11.0. This was chosen to meet the pull away targets on a 30 % grade at gross vehicle weight. The transmission features a standard wet sump lubrication.

The results show the energy consumption for various driving cycles with a maximum vehicle speed limited to 100 km/h, in particular the previously illustrated Stuttgart Cycle. The speed limit is defined by the requirement for short-distance travel on highways or A-roads, i.e. driving from the city center to an airport on the outskirts. The results of the full vehicle simulation are shown in Figure 5, with both the maximum electrical range and the energy consumption within the drive cycles NEDC, WLTC and the urban Stuttgart Cycle listed. The speeds over time profiles also clearly reflect the maximum velocity limit of the MEET vehicle.

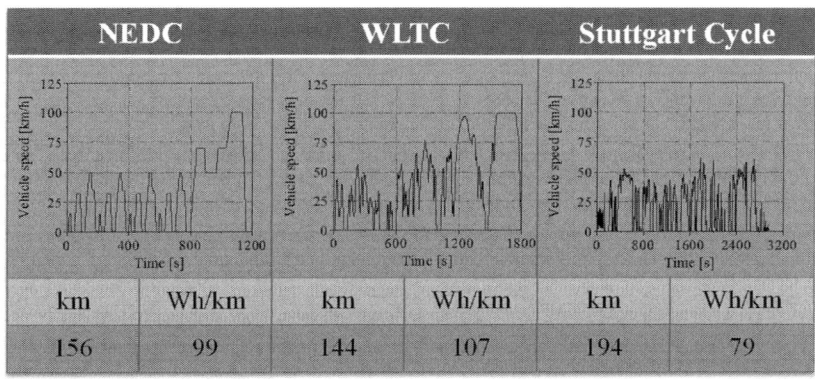

Figure 5: MEETs electrical range and energy consumption in the NEDC, WLTC and Stuttgart Cycle

Based on the simulations the electrical range within characteristic urban driving conditions, represented by the Stuttgart Cycle, is up to 194 km. The energy consumption of the MEET thereby comes down to 7.9 kWh per 100 km.

Further vehicle simulation show the driving dynamics of the MEET. This is presented in Figure 6. It shows both the distance travelled and maximum speed after 1, 3 and 5 seconds.

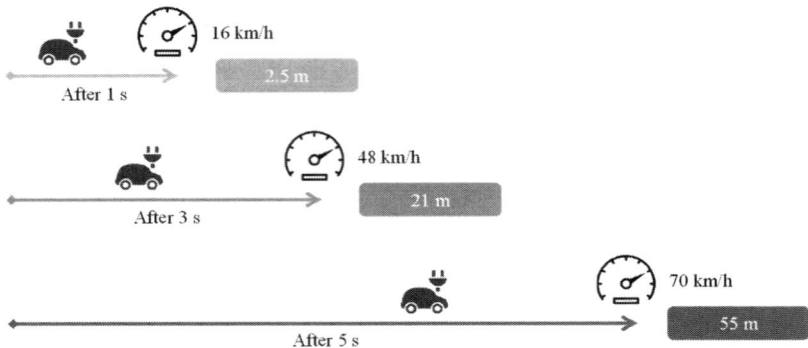

Figure 6: MEETs vehicle dynamics

4.2 Thermal management

It is a well-known fact that the energy consumption of auxiliary devices substantially compromises the drive range of electric vehicles. One of the main contributors to additional energy consumption are the vehicles' air conditioning systems – for cooling the cabin in summer and heating in winter. User experiences as well as test and simulation data indicate that the drive range at winter ambient temperatures can decrease by 30 to 50% compared to the nominal values [12]. For a vehicle used mainly in urban areas – like MEET – the influence of cabin heating on the drive range is very strong if traditional heating methods, for example a PTC-heater were used. This is because the energy consumption for moving the vehicle is comparably low due to the low mass of the vehicle and the low average driving velocities in urban situations. At the same time the energy consumption for heating the cabin would remain high because it does not depend on these two parameters. In a drive cycle which represents well the use profile MEET is designed for – like MAHLE's Stuttgart cycle, an urban drive cycle with an average speed of 21 km/h – the energy consumption for heating may be even higher than for driving.

In order to minimize the energy consumption for heating MAHLE has chosen a different concept other than warming up the cabin air flow by a resistive heating element like a PTC. This concept is mainly based on two approaches. One is the reduction of the energy amount required to establish passenger thermal comfort by the use of localized radiative heating. The second is the reduction of the energy amount to warm up the remaining necessary cabin air flow by the consequent use of available waste heat at high efficiency through a thermoelectric coolant to coolant heat pump.

To reduce the energy requirement for thermal comfort thin heating elements were integrated into several surfaces of the vehicle's interior functioning as radiative heaters as shown in Figure 7.

They allow a very flexible and fast responding generation of radiative heat only when and where it is needed, taking into account the number and location of occupants inside the vehicle and their individual thermal needs. The radiative heat transferred to the passengers can compensate a lower cabin air temperature which subsequently reduces heating energy demand [13].

Figure 7: Surface heating within the MEETs cabin

In addition to the radiative heat transfer a minimum flow of warmed-up air is still required to maintain mist- and frost-free windows, necessary cabin air exchange and to support passenger thermal comfort. To minimize the energy amount required for the warm up this air flow a thermoelectric heat pump (Figure 8) is used in the MEET concept to make the waste heat of electric powertrain components available for this purpose. Compared to refrigerant compression heat pumps this thermoelectric (TE) device has several advantages for the use in the MEET vehicle. It is a solid state component without moving parts that can pump available waste heat to a useable temperature level at a coefficient of performance of around 2. Moreover, its efficiency can be adjusted to lower values by control means which may sound illogical for a requirement but is an important feature of this device. In urban traffic situations the availability of waste heat from MEET's drive train components is quite limited and usually does not exceed several hundred Watts. This makes an additional heating device necessary to cover the remaining heating requirement. In refrigerant compression heat pump systems usually a separate PTC heater is used, making the heating system more complex. TE-heat pump systems do not need any extra heater components since they can produce the required amount of heat internally while still pumping the drive train's waste heat to the required temperature level. In specific cases the thermal inertia of the vehicle's battery can also be used as a storage heat source for the TE heat pump.

Figure 8: Thermoelectric heat pump from MAHLE

Together with the air conditioning system using eco-friendly CO_2 (R744) as refrigerant MEET combines smart new technologies to a thermal management and passenger comfort system at the most efficient level.

The benefit of this combination of technologies on the vehicle's drive range is shown in Figure 9 at an ambient temperature of 0 °C. Especially in the drive cycle with the highest urban usage – Stuttgart cycle – it outperforms PTC heating solutions by more than 50 % of extra drive range.

12

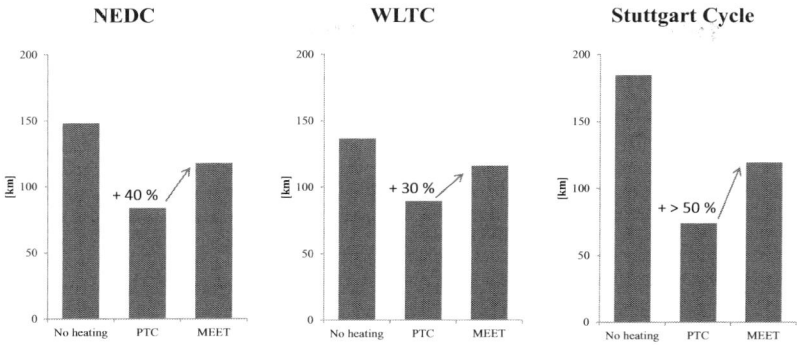

Figure 9: Influence of MEETs thermal management technologies on the electrical driving range

5 Summary and Conclusions

The current challenges for the automotive industry as well as the advancing urbanization lead to new mobility concepts and powertrain technologies. MAHLE developed a fully electric urban vehicle concept called MEET – MAHLE Efficient Electric Transport. It combines MAHLE's core competence and expertise in electric powertrains, thermal management, as well as climate comfort. Right at the start of the development key aspects of urban mobility were identified to determine the characteristics of MAHLE's urban vehicle concept. MEET gives an outlook for future mobility and it shows the holistic approach for an efficient system solution for the electrified urban mobility.

MEET's electric powertrain consists of two 48 V motors with a central gearbox. Each motor delivers 30 kW peak power and provides high peak efficiency of up to 97 % within urban traffic driving (94 % in combination with the power electronics). Traction motor and power electronic are developed in one assembly to maximize the flexibility in terms of layout and packaging. The 48 voltage level offers significant potential in decreasing cost compared to a high-voltage powertrain with equal power output and increases the usability for maintenance. Within a typical urban drive cycle the energy consumption of the MEET vehicle is 7.9 kWh/100 km. This results in an electric driving range of 194 km.

The electric driving range is an important factor for customer acceptance. Due to the higher required heating and cooling power within cold and hot outside conditions the driving range can decrease by 30 to 50 % compared to nominal values. MAHLE developed new technologies to a thermal management and passenger comfort system at the most efficient level to increase the vehicle's driving range.

To validate the simulation results vehicle testing is scheduled for the second half of 2018.

6 References

[1]   United Nations Framework Convention on Climate Change: The Paris
      Agreement, Online verfügbar, http://unfccc.int/paris_agreement/items/9485.php.
[2]   IPPC, 2014: Climate Change 2014: Synthesis Report. Contribution of Working
      Groups I, II and III to the Fifth Assessment Report of the Intergovernmental
      Panel on Climate Change [Core Writing Team, R.K. Pachauri and L.A. Meyer
      (eds.)]. IPCC, Geneva, Switzerland, 151 pp. page 47
[3]   VDA: CO_2-Regulierung bei Pkw und leichten Nutzfahrzeugen, Online
      verfügbar, https://www.vda.de/de/themen/umwelt-und-klima/co2-regulierung-
      bei-pkw-und-leichten-nfz/co2-regulierung-bei-pkw-und-leichten-
      nutzfahrzeugen.html, 2017.
[4]   Thomson Reuters: Paris will bis 2030 Benzin- und Dieselautos verbannen,
      Online verfügbar,
      http://www.handelsblatt.com/unternehmen/industrie/frankreich-paris-will-bis-
      2030-benzin-und-dieselautos-verbannen/20446350.html, 2017.
[5]   Volkery, Carsten: Autofahrer müssen „Vergiftungsabgabe" zahlen, Online
      verfügbar, http://www.wiwo.de/politik/ausland/london-autofahrer-muessen-
      vergiftungsabgabe-zahlen/20487480.html, 2017.
[6]   Novak, Marja: Slovenia to ban new fossil-fuel cars from 2030, reduce debt,
      Reuters, Online verfügbar, https://www.reuters.com/article/slovenia-
      autos/slovenia-to-ban-new-fossil-fuel-cars-from-2030-reduce-debt-
      idUSL8N1MN54J, 2017.
[7]   Pieters, Janene: New dutch government's plans for the coming years, Online
      verfügbar, https://nltimes.nl/2017/10/10/new-dutch-governments-plans-coming-
      years, 2017.
[8]   Dameri, Renata P.: Smart City Implementation: Creating Economic and Public
      Value in Innovative Urban Systems, Springer International Publishing, Cham,
      2017.
[9]   Fischer, Wolfgang: Intermodalität, e-mobil BW, Online verfügbar,
      http://www.e-mobilbw.de/de/innovative-mobilitaet/intermodalitaet.html.
[10] Fritsch, Karl-Martin; Schmülling, Christoph; Wieske, Peter: Designed by Power
      Demand: An Electric Drive System for Urban Mobility, 26th Aachen
      Colloquium Automobile and Engine Technology 2017, Stuttgart, 2017.
[11] Fritsch, Karl-Martin; Christoph, Schmuelling; Tobias, Binder; Markus,
      Cramme: Challenges of a 48 V Drive System with 20 kW continuous
      mechanical power, ELIV 2017, Stuttgart, 2017.
[12] Leighton, Daniel: Combined Fluid Loop Thermal Management for Electric
      Drive Vehicle Range Improvement, SAE Int. J. Passeng. Cars – Mech. Syst,
      vol. 8, no. 2, 2015.

[13] Schmidt, Carolin; Veselá, Stephanie; Bidhendi, Mariam N.; Rudnick, Jana; van Treeck, Christoph: Zusammenhang zwischen lokalem und globalem Behaglichkeitsempfinden: Untersuchung des Kombinationseffektes von Sitzheizung und Strahlungswärmeübertragung zur energieeffizienten Fahrzeugklimatisierung in FAT-Schriftenreihe 272, Verband für Automobilindustrie, Berlin (Hrsg.), Lehrstuhl für Energieeffizientes Bauen, RWTH Aachen, 2015.

AllCharge™ – a user-centric solution for traction and charging

Friedrich Graf, Dr. Martin Brüll, Siegmund Deinhard

Continental, Regensburg, Germany

© Springer Fachmedien Wiesbaden GmbH, ein Teil von Springer Nature 2018

J. Liebl (Hrsg.), *Der Antrieb von morgen 2018*, Proceedings,

https://doi.org/10.1007/978-3-658-21419-7_3

Abstract

The key purpose of the powertrain of an electric vehicle is to provide adequate driving power and sufficient energy to fulfil the range request of the driver. The question when, where and how to charge the battery gets increasingly in focus. Maximizing the degree of freedom to charge compatible to different AC and DC power levels at any charging point is an important step to strengthen the attractiveness of electric vehicles. Furthermore vehicles with 60–100 kWh battery capacity can only be operated with appropriate high charging power (> 100 kW) at longer distances. Nevertheless it´s also important to provide highest charging efficiency for the majority of typical charging events at 3 to 11 kW. A useful functionality of an electric vehicle is also to provide 230 V AC electric power to devices anywhere (V2D). Just as well electric power of the battery can be used to shave peak loads of the power grid applying (V2G). This integration contributes to smart grid features but also requires compatibility of an electric vehicle to classical electric installation at optimal cost. The Continental AllCharge System offers a forward-looking technology.

Introduction

The population cites especially long charging times, the low density of charging points and incompatibilities of different systems, voltage and power classes as important obstacles for the further spread of purely electric vehicles. Continental paid attention to this challenge: A unique, so-called AllCharge system has been developed which eliminates all the aforementioned barriers, thus contributing decisively to the breakthrough of electric mobility and also offers a number of additional, far-reaching benefits for the future sustainable energy supply.

The much-discussed 'range anxiety' in electric mobility should no longer play a major role from 2020 onwards. In addition to larger batteries, the faster and more comfortable charging options will contribute to overcome this. Continental has developed the innovative AllCharge system, to enable car drivers to use any charging station – regardless of the charging power, current type and voltage level. Instead of packing more charging technology into the car, the engineers evolved the electric propulsion into the charging system. Thus the electric motor and the inverter (converter between DC and AC) have been upgraded to accomplish the additional task of charging. As the only additional component, this system includes a DC/DC converter, which ensures an always optimal voltage to the battery. While charging either alternating current flows from the charging station through the electric motor to the inverter and from there as a direct current to the battery; or the charging station releases DC power which directly charges the battery via the DC/DC converter.

2

Figure 1: AllCharge makes EVs fit for any type of charging stations

AllCharge gives the driver full flexibility in wired charging and has access to all charging opportunities offered by the infrastructure. While AC charging, the power limitation is eliminated due to the on-board charger. If the AC infrastructure allows it, the AllCharge can charge the battery with up to 43 kilowatts, this means up to 40 kilometers in ten minutes. The new Continental system also makes it possible to use 400-volt DC fast charging stations, which can reach up to 150 km in ten minutes. Premium vehicles with a very large battery can even increase their range at 800-volt DC stations by up to 300 kilometers in ten minutes. Thereby, the charging time approaches significantly to the duration of stay at a conventional filling station. With its *vehicle-2-device* technology, the AllCharge system can charge mobile electrical devices (e.g.: laptop, refrigerator or electric drill) from the vehicle battery.

AllCharge Description

AllCharge Topology

In the case of wired charging of an electric vehicle, the alternating current (AC) from the grid must be converted to direct current (DC) for charging the battery. This can be done on the one hand in the charging station, which then passes DC power to the car: the so-called DC charging. DC fast charging is required to derogate range anxiety and allows long-distance EV travel along the traffic main streams.

On the other hand, AC can be led into the vehicle and the conversion would be done on board with the help of a so-called on-board charger (OBC): the so-called AC charging.

As DC charging stations are disproportionately more expensive than AC charging stations, usually the AC charging stations are used for daily needs and the DC charging focuses on the highways (about every 50 km) in support of the long distance traffic [1], [2], [3]. With rising battery capacities and growing electric car pools, the demand for even higher charging power will increase. In order to make this possible in everyday life, from a societal point of view the AC charging capacity of electric cars should therefore be significantly increased [4].

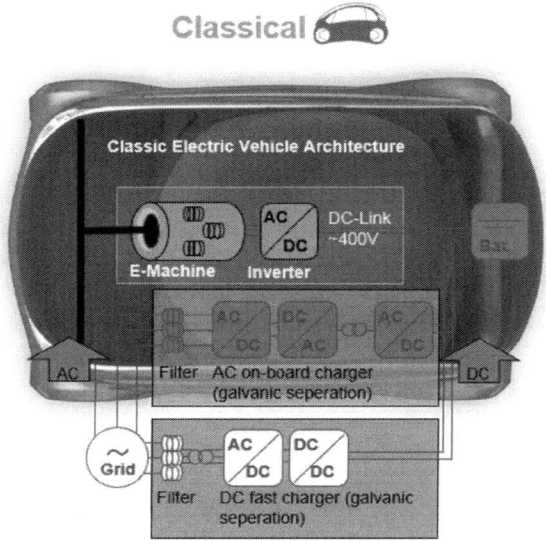

Figure 2: Classical EV HV architecture topology: Drivetrain & On-board charger separated

The obvious solution according to the classic architecture of electric vehicles would be an increase in performance of the OBC. This leads to more space requirements and rising costs and is one of the reasons why the OBC today typically supports only 3.7 kW (about 0.3 km/min) charging power. Continental has developed a solution with All-Charge to enable AC charging without a dedicated OBC. This is achieved by converting the propulsion to a charger. Thus, a very powerful component is reused on board and the bottleneck of the OBC, resulted of economic reasons, removed from the vehicle. It is therefore possible to display AC charging powers of 43 kW (approximately 4 km / min) almost everywhere in urban areas [4].

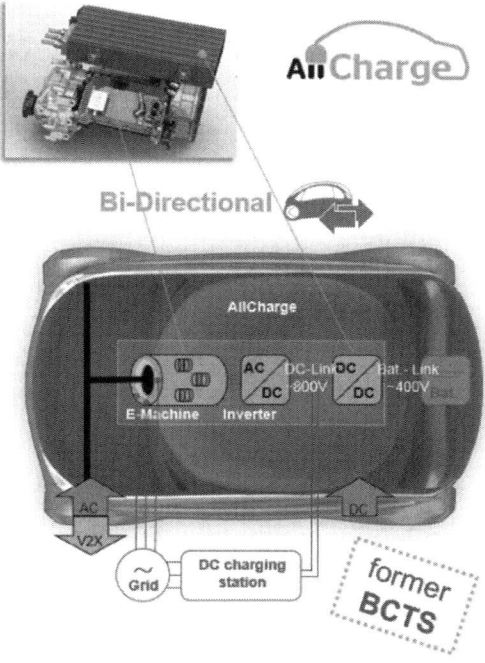

Figure 3: AllCharge topology

The AllCharge system also includes a DC-Booster, which represents the DC voltage adaption between the propulsion and the battery. In addition to the AC charging function, this makes it possible to set a variable intermediate circuit voltage in case of traction and therefore to use the drive significantly more efficiently. Simulations show a gain of system efficiency of at least 3.5% for traction over classical architectures. Due to the intermediate circuit voltage the drive can always be operated at an optimum operating point. The increased range results in a direct elevation of system value compared to a classic architecture. Furthermore, it accelerates DC fast charging by up to 10% and not only by the current 400V but also in future 800V charging stations. This effect is achieved by the fact that – as with classic DC charging systems – the charging station and the connection to the vehicle represents a current limit, but the voltage depends no longer on the state of charge (SOC). In classical architectures, the charging power is therefore limited to 75% of the maximum for an empty battery. The charging power (product of electricity and voltage) can therefore be fully utilized from the beginning when using the DC Booster of the AllCharge system.

In contrast to common OBC technologies, all components are designed for bidirectional use. Thus, the battery energy can also be easily used for new applications than for traction. AllCharge offers the possibility to deliver AC directly to the house. Consequently, this system saves on an innovative way that additional inverter and therefore considerable costs. A dedicated home battery storage can also be omitted, as the electric vehicle takes over this role. Furthermore, a household socket can be operated directly on the vehicle without a further conversion component. This feature allows the AllCharge system to operate electrical equipment, such as a laptop, refrigerator, or drill, regardless of a power grid availability.

Figure 4: AllCharge demo car providing AC to a fridge

Unique feature of AllCharge

With AllCharge, Continental has developed the only system in the world that supports all current and future conductive charging modes for electric vehicles: Single and three-phase AC charging at all voltages (120 V / 240 V) and frequencies (50 Hz / 60 Hz), DC charging at all current and expected voltages (400 V / 800 V). In addition, all types of charging are shown with the maximum possible power with high efficiency and therefore with the highest user benefit due to maximal flexibility.

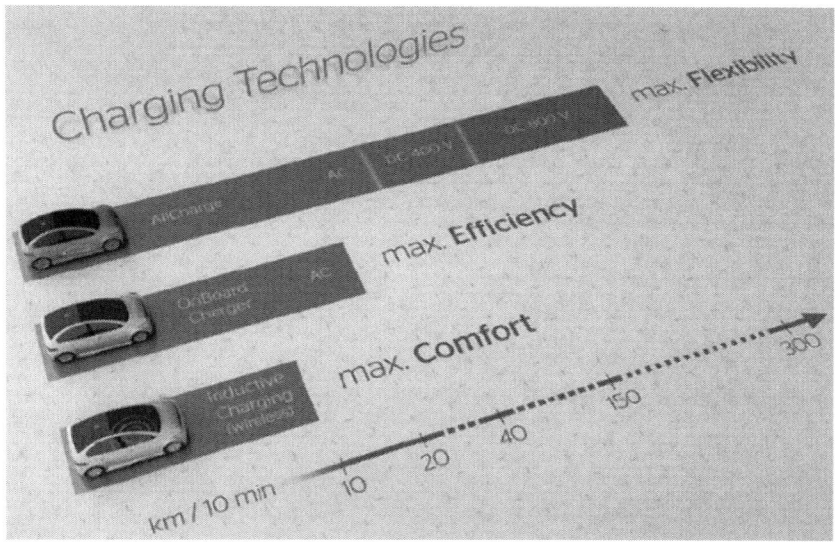

Figure 5: Comparison of three charging techniques: Inductive Charging, classical On- board charger and AllCharge

Use cases for Charging and Vehicle-2-X

The integration of the charging functionality into the drivetrain has multiple functional benefits. It enables additional charging functions, optimizes the traction efficiency and opens the door for innovative bidirectional *vehicle-2-x* functionalities and saves cost. In this paper will focus on use cases for *charging* and *vehicle-2-x*.

Charging Use Cases

For space- and cost reasons, most of current e-vehicles do not have high power charging devices. The majority of these vehicles are charged at AC charging stations with 3.5-7.2 kW, in the next vehicle generation with 11 kW. On the other hand, the current charging infrastructure often offers AC fast-charging systems already with outputs in the range of 22-43 kW (up to 43 kW for the 3-phase system in Europe, or around 20 kW worldwide). In addition, DC charging infrastructure is in the roll- out phase, which will provide charging power up to 350 kW. The illustration in Fig. 6 summarizes the available power classes in different charging situations in the lead markets of Europe, USA and China.

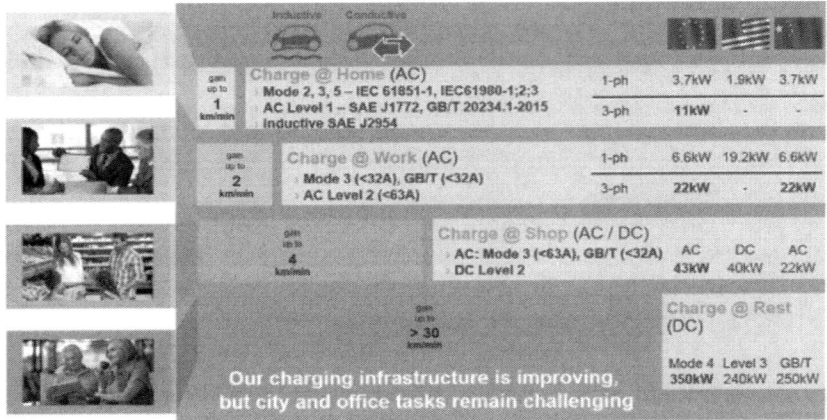

Figure 6: Charging use cases vs. maximal infrastructure capabilities in the three dominant EV markets

In this analysis, we focus on the end-user's application. For this purpose, we consider charging use cases, which are already applied in practice and the future *vehicle-2-x* applications. Technically, a charging power is discussed in the unit of *kW*. The end-user's view is more interested in the unit of *km/min and cost/km*. The conversion requires an average energy consumption of the vehicle. To meet a good average value for C-segment cars in real drive, we assume a consumption of ~180 Wh/km. This leads to following rule of thumb:

11 kW charging power corresponds to **1 km/min** charged range

Another aspect is the *cost for charging* for the end-user. [5] has set the cost for electrical energy for 4 years operation period for a Germany vehicle with 15.000 km p.a.

0.28 €/kWh

to reach roughly same cost/ km as the comparable combustion candidate.

So access to inexpensive electricity will play a major role in most of the buying decisions. To reach tariff flexibility, high (AC) charging power necessary: The widespread time- based pricing at public charging stations e.g. 4 Ct./min. would lead for 7kW charging power to 34 Ct/ kWh. Even if the battery is full, the clock continues to run. In simple words, double the charging power means half the cost, not included fix cost components as roaming- and parking fee. For occasional quick charging eg. *charge@rest*, higher energy cost would be accepted [12].

The third fundamental aspect for a use case analysis is the *availability of charging options* for the particular user. Following a survey [12], the availability of 22 kW AC high power charging is regarded as an important pillar and to cover wider ex- urban rural areas. High power charging > 50 kW is seen as 2nd pillar in the future.

Use cases such as car sharing, automated car fleets on public or restricted areas are not considered here, but inherit other specific business cases and resulting charging scenarios and are driven by cost of ownership.

Employees with access to *charge@work* are six times more likely to purchase an electric vehicle [6]. Therefore, as also stated in [7], *charge@work* is a clear market trend and discussed at companies and governments to increase attractiveness and market penetration of EVs.

Estimating a typical driving distance of 15.000 km per year with 180 Wh/km and a charger system efficiency of 90% the vehicle needs to charge 3000 kWh per year. Doing that with 3.7 kW would last for 811 hours per year.

Since charging incorporates the battery, an estimation of the charging acceleration with higher charging power cannot be done with rule of three. Based on measurements of series vehicles, we propose following model. The maximal charging power of a battery temperature depends on the C- rate, which is the factor power/capacity. The C- rate today as state-of the art is mostly limited at 2 for all mass production cars (when optimized for highest energy density). An additional limitation comes via the battery's state-of-charge (SOC): 0.1C at 0-5 % and 95-100 % SOC, 0.5C at 5-15 % and 90-95 % SOC, 1C at 80-90 % SOC. Temperature dependencies are not regarded here. Assuming further a 50 kWh battery for a typical future mass production car, we come to an effective average charging power over all SOC values like shown in Tab. 1. This effective charging power is considering 0 % to 100 % charge with reduced C-rate at beginning and end of the charging process as shown in Fig. 7.

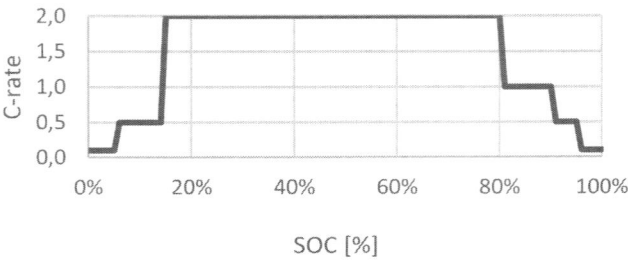

Figure 7: Charging C-rate over SOC for state of the art C-class vehicle from measurements at room temp.

Table 1: Effective average charging power for 0-100% SOC charging (from examples)

Charging station [kW]	Effective charging power [kW]
3.7	3.7
11.0	10.4
22.0	20.3
43.0	36.5
150.0	74.3

In essence, more charging power does mean proportional reduction of time, but requires higher performance of battery cell technology. Nevertheless, *charge@rest* needs high power to cover long range driving and assumes such more advanced cell technologies.

It is found that the majority of charging is *charge@home* according to [8] 60 % to 80 % at home. This will be additionally driven by tariffs that incentivize at-home, off-peak charging (as compared to daytime charging).

The second important charging use case is workplace *charge@work* with 30 % to 40 % [12]. Public charging, like DC fast charging is only minor used 3-4 % as solution for longer range driving and if no charging at home was available. Our investigation of today's installed infrastructure (see chapter 2.4) often showed the use case *charge@shop* as motivation. Taking all together and the total mileage resulting from [6], we derive the use case occurences, like shown in Tab. 2.

Table 2: Power levels for charging

	Occurrence	Effective Power [kW]	Hours/year
Charge @ Home	55%	10.4	159
Charge @ Work	30%	20.3	44
Charge @ Shop	10%	36.5	8
Charge @ Rest	5%	74.3	2

This model results in a charging time per year of only 213 hours, which is just 26 % of the 811 hours of a pure 3.7 kW charging system. A typical car with DC fast charging and only 3.7 kW AC charging would still require 772 hours. A driver of such a car got the promise of a fast charging car, but charging takes effectively 3.6 times longer than with a system supporting also fast AC charging. We call this phenomenon the *DC charging trap*.

Furthermore, the fast AC charging can be regarded as for *vehicle-2-x* functions, since the gained flexibility in charging is mandatory for shifting demands. On the opposite it can be stated, that with slow AC charging power in the vehicle, only little *vehicle-2-x* is possible, since the vehicle has often to be charged at maximum available charging power in order to keep up availability of the vehicle [6].

Vehicle-2-X Use Cases

In addition to the classic use cases of *propulsion* and *charging* of an electric vehicle, new applications are being developed, which a combustion vehicle cannot support, because due to the large battery, new possibilities for mobile energy supply have been opened up: the various *vehicle-2-x (V2X)* applications. In the *vehicle-2-device*, household electrical devices such as e.g. laptops, refrigerators or tools such as drills powered by a power socket located on the vehicle. In the US., supplying external tools and machines with the power of a car has a longer history. AllCharge as 'power bank' could make the electrified version popular.

In *vehicle-2-home*, the vehicle supplies a residential building with electrical energy. In Japan *vehicle-2-home* is standardized for emergency- and grid balance reasons. This technology was developed in the wake of the loss of nuclear energy after the Fukushima incident notably in Japan, the electric cars should support households with unstable network. Nissan developed the "Leaf-2-Home" system, which uses the DC charging interface to provide AC power to the house via a converter similar to a photovoltaic inverter [9]. Extensions of the VDE 0100-551 (direct coupling of different electric sources in house- or public grid, switched parallel) will pave the way.

Figure 8: Use case *vehicle-2-device* in action

At *vehicle-2-site*, a factory site uses the energy stores of the employee vehicles to cushion the peak power requirements and therefore to reduce electricity costs – a good combination with with *charge@work*. In the *vehicle-2-grid*, the energy provider of the car pool makes use of e.g. public charging stations to ensure frequency stability.

Often, these use cases are associated with the charging applications. It turns out that the V2X use cases in particular require AC-regeneration and thus emphasize the need for AC-based bidirectional charging solutions. Although these applications support the success of the energy transition, the market has so far taken up on only a very limited part of the possibilities of mobile energy supply.

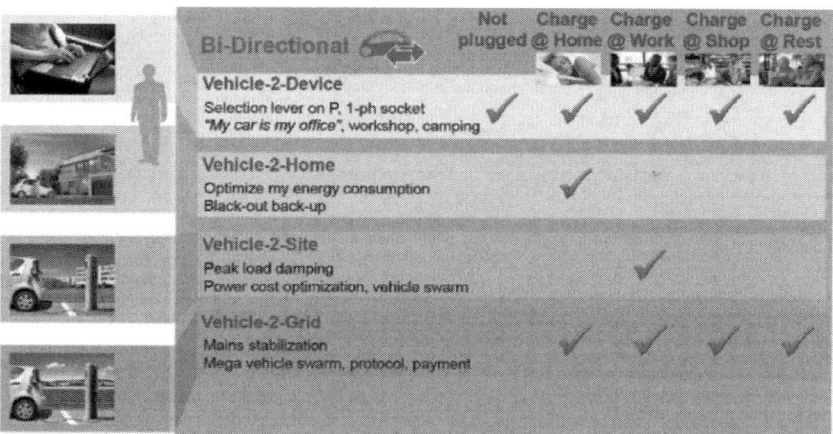

Figure 9: Vehicle-2-X use cases in correlation to Charging use cases

In Germany a fundamental change of the (electric) energy market is predicted. A more de- centralized generation, transport, supply assumes storage capacities and IT infra-structure. The obstacles are not only technical ones, but refer to regulation as well (eg. taxation) [5].

Finally, a de-centralized organization of the grid and related *vehicle-2-grid* functions depend on the attractiveness of this use case, e.g. less costs for energy if the storage function is enabled by the owner on an EV.

So the *vehicle-2-x* use cases represent attractive further options for now and in the fu-ture – that's why AllCharge incorporates these.

To keep in mind, the buying decision for a specific vehicle with specific charging is not only a strictly rationale decision. Vehicle owners regard flexibility as value of its own, that's why allCharge is attractive, too from this viewpoint.

Infrastructure

The regulation-driven automotive industry is pushing towards zero emission from the tank to wheel and therefore towards e-mobility. However, the economic issues of the infrastructure must also be taken into consideration, which may lead to demands on the vehicle.

Home Charging

Electricity is everywhere and highly flexible use. So it's obvious to simply use it also for transportation, at least for the daily commuter distance. Comparing the energy consumption of a house and a vehicle, one finds out how energy consuming transportation is. A German 4-persons household has a yearly consumption of 4200 kWh [10]. Estimating a typical driving distance of 15.000 km per year with 180 Wh/km and a charger system efficiency of 90 % a vehicle needs to charge 3000 kWh per year. So transportation consumes as much as the magnitude of household electricity.

Depending on the worldwide regions, home charging at about 2-3kW is often possible and sufficient for the commuter's needs. Overnight, when the household consumption is very low, the remaining power capacity could be used to charge the car with. Usually the batteries tends to be at high SOC in the morning. Capacity bottlenecks in a residential area should be solved via smart charging solutions to avoid charging peaks in the evening.

Home Charging is the backbone of the charging infrastructure.

Public Charging

Beside Home Charging the Public Charging infrastructure is mandatory. The majority of the population does not live in a detached house with Home Charging capability. So with growing EV volume, more and more drivers depend on alternative charging solutions. So proper Public Charging is key.

Such a user needs nearby solutions. But since the parking time cannot be as long as at Home Charging, higher charging power is required. The same request comes from drivers probably having the chance for Home Charging, but are e.g. on an inter-city trip and need to recharge some amount nearby the current halt. Furthermore the use case *charge@shop* (see Fig. 6) has similar power demands, often in urban regions.

So why not simply solving this need with DC fast charging infrastructure like on the motorway?

Figure 10: Today's Motorway infrastructure: A fast charging station each 50 km along the motorway

Having DC fast charging at the motorway is the best solution for this use case *charge@rest*: On a long way trip, any minute counts. So charging should be as fast as possible and this can be provided only with DC charging stations. The first proprietary network of such kind was implemented by Tesla. The more general approach is following nowadays supporting the CCS, CHAdeMO or GB/T standards (Tesla, SLAM [11], Tank & Rast, Ionity). The investment is mainly covered by governmental funding and leads to a charging station density of one per about 50 km (see Fig. 10).

The request on consumer side, following the nearby request as described above, lead to an average density of fast charging stations comparable with today's petrol station density on average. This guides us to a field density of one every 5 km. Fig. 11 indicates that such an approach need 100 times more fast charging stations like already planned.

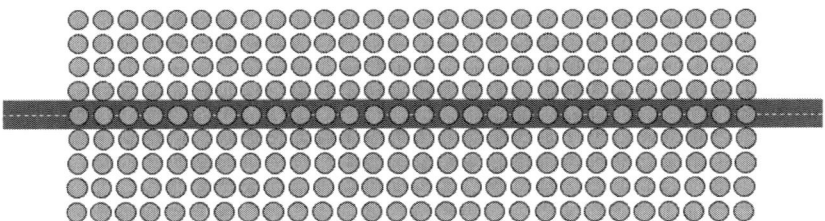

Figure 11: Needed Field infrastructure: A fast charging station every 5 km in the area

In [7] it is explained that "AC Charging is the dominant type of plug-in vehicle charging and will remain that way into the long term. This is because domestic charging is the domain of AC Charging. It is a cheap and convenient way of charging. AC will also play an important role in the public domain also." This is reflected in a forecast in market numbers that […]" globally by the end of 2016 there were just under 3 million AC charging stations cumulatively. This is set to rise to over 32 million by 2023. Furthermore, [7] estimates that in 2016 there were 3,300 DC charging stations cumulatively, which is set to rise to over 62,000 by 2023." So the total amount of AC charging stations will be a factor of ~500 higher than DC charging stations and this is not taking the home charging possibilities into account. This difference in numbers is driven by the higher cost of the DC stations, which can be as low as 370$ for a level 2 charger and more than 45.000 $ for a DC charging station [7]. This deviation is even higher than former publications indicated [12]. It can be explained by increasing charging power demand in the last years, which dominate the DC charging station costs. And, as mentioned above [12], the availability of 22 kW AC high power charging is regarded to secure the density in wider ex- urban rural areas.

Since the public charging technology is and will be diverse, this requires on vehicle side a flexibility as AllCharge provides to be sure getting proper use of the installed charging stations.

Summary

E- mobility is an important element in achieving fleet emissions targets. To achieve the required market volumes, a user-centric solution is necessary which puts the user in the center and weights charging equivalent to traction. In this paper we have analyzed a variety of Use Cases and cross checked it with the current and future infrastructure. Each driver will live a more or less individual combination of these use cases and this could temporarily also change, e.g. for a holiday trip and for rental cars. The infrastructure provides a variety of different charging techniques and this will not change in the next years. So buyers will decide for vehicles providing maximal flexibility towards the drivers charging source. Bidirectional operation, starting with *vehicle-2-device* as new use case, serves further needs of the customers and might enable new busines cases. This evolution will be continued with emerging *vehicle-2-home* and *vehicle-2-grid* use cases. The presented AllCharge system provides this maximal flexibility towards the charging use cases and further supports all the future *vehicle-2-x* use cases.

AllCharge could not only accelerate the previously hesitant market launch of electro-mobility through its high level of customer acceptance, but also promotes due to its additional features the success of the energy 'revolution' as a whole.

References

[1] M. Bruell, P. Brockerhoff, F. Pfeilschifter, H.-P. Feustel and W. Hackmann, „Bidirectional Charge-and Traction-System," in *EVS29 Symposium*, Québec, Canada, June 19-22, 2016.

[2] J. S. Johansen, *Fast-Charging Electric Vehicles using AC,* Technical University of Denmark, 2013.

[3] Graham Evans, „Electric Vehicle Charging Infrastructure: Definitions and Market Analysis," IHS Markit, March 2017.

[4] „Combined Charging System Specification," [Online]. Available: www.charinev.org. [Zugriff am 15. 12. 2017].

[5] „Wie rentabel sind Elektroautos?," [Online]. Available: https://www.adac.de/infotestrat/adac-im-einsatz/motorwelt/e_auto_kostenvergleich.aspx. [Zugriff am 15. 12. 2017].

[6] M. Brüll, „Bidirectional Charge and Traction System: A convenient solution for E-Mobility," in *Future Powertrain Conference*, Birmingham, UK , 2017.

[7] M. Brüll and P. Brockerhoff, „Avoid the DC charging trap – high power everywhere charging," in *EVS30 Symposium*, Stuttgart, Germany, October 9-11, 2017.

[8] O. Maiwald and M. Brüll, „Bidirectional Charge and Traction System (BCTS) – A combined Solution for Electromobility and Charging," in *Distributive Technologies Conference*, Santa Clara, USA, 2016.

[9] K. Ogawa, "Vehicle-to-Home technology boosting the value of automobiles," Nikkei BP Japan Technology Report / A1407-060-011, 2014.

[10] „Stromverbrauch im 4-Personen-Haushalt," 15. 12. 2017. [Online]. Available: https://www.die-stromsparinitiative.de/stromkosten/stromverbrauch-pro-haushalt/4-personen-haushalt/index.html.

[11] D. Horn, A. Bauer, A. Schmidt and O. Udovenko, „The influence of investment expenditures on the development of fast charging infrastructure," in *EVS30 Symposium*, Stuttgart, Germany, 2017.

[12] M. Brüll, P. Brockerhoff and M. Töns, „Bidirectional Charge and Traction System: Solution for integrated high power AC charging," in *APE Conference*, Paris, France, 2017.

[13] S. Elektromobilität, „Bedarfsorinetierte Ladeinfrastruktur aus Kundensicht http://schaufenster-elektromobilitaet.org/de/content/dokumente/dokumente_1/dokumente_2.html," 2016.

SOFC EV range extender systems for biofuels

Juergen Rechberger, Michael Reissig, Vincent Lawlor,
AVL List GmbH

© Springer Fachmedien Wiesbaden GmbH, ein Teil von Springer Nature 2018
J. Liebl (Hrsg.), *Der Antrieb von morgen 2018*, Proceedings,
https://doi.org/10.1007/978-3-658-21419-7_4

The automotive industry is heavily investing in Battery Electrical Vehicles (BEV) and to a far lesser extend in Fuel Cell Electrical Vehicles (FCEV). BEVs provide the highest overall energy efficiency but require very large and expensive batteries for long driving ranges. Additionally the recharge time of batteries remains a concern, as quick charging will be limited by electrical grid stability requirements. For FCEVs mainstream technology in the industry is PEM based on Hydrogen. Due to the additional energy conversion, Fuel Cell is in overall less efficient than BEV but provides advantages towards driving range and refill time. Nevertheless, the hydrogen station rollout is slow and limits the market penetration of these vehicles.

To be independent from hydrogen infrastructure, various different fuels (ethanol, methanol, …) for fuel cell vehicles have been explored by various OEMs. However, all these efforts have been stopped due to the increased complexity and high cost. Since roughly 2 years a consortium consisting of NISSAN, AVL and Plansee is exploring a completely new approach based on SOFC (Solid Oxide Fuel Cell). This fuel cell comes from stationary applications and is very fuel flexible. Due to the high operating temperatures (600-800°C) it can tolerate very high CO levels and therefore it can operate more or less on every kind of reformed fuel input. NISSAN presented in 2016 a first Concept Car called e-Bio Fuel Cell during the Olympic Games in Brasil. This vehicle is based on the ENV200 with a SOFC Range Extender of 5kW. This range extender is operated by pure Bio-Ethanol and recharges the HV battery. By this approach, the driving range of the vehicle could be extended from 150km to 600km with a 30L ethanol tank. In two follow-up Joint Research Projects (MestRex & COMPASS) this consortium is now focusing on to develop this technology towards product maturity in the timeframe 2022-24. Within these projects completely new stack technology is developed based on metal supported cells. In these cells most of the ceramic material is replaced and eliminated by a porous metallic substrate. This new technology approach will enable quick start in less than 15min and also lower operating temperatures.

With SOFC Rang extenders the driving range of BEVs can be extended. The process will not be completely zero emission, but will have extremely low emissions below e.g. 1ppm NOx. If biofuels are used, the carbon circle can be closed. The tailpipe CO_2 emissions are significantly lower compared to engine or hybrid vehicles as the efficiency of the SOFC range extender is in the range of 50-60%.

Targets of MestRex and COMPASS projects

In Table 1 performance and physical targets for the MESTREX and COMPASS systems are shown, alongside the long term AVL system targets. The requirement in COMPASS for a 15 minute start time is a signifigant challenge. Furthermore, the additional system cost and weight/volume requirements will force the BEV-SOFC-REX system design in a direction even more suitable for the automotive industry. For the rapid start some novel methods like sending combusted gas into the cathode and/or anode channels is being investigated. The key concern is related to the SOFC cathode electrode material, which may degrade unacceptably over the system lifetime. Other methods in order to drive heat rapidly into the stacks, during start up, are also being investigated as part of a multi level approach. In this approach several techniques may be used as a function of system temperature in order to rapidly heat the SOFC stacks.

The MESTREX system will use a state of the art heat exchanger in order to transfer heat from a start burner to the stack air supply. A 30 minute start with a single HEX is on the limit for a rapid start considering the temperature restrictions within such a system. The noise level requirments will also be quite a challenge to reduce. Special solutions for the system air blower and exhaust gase silencer will need to be investigated.

Table 1: MESTREX, COMPASS and AVL eventual targets

	MESTREX	**COMPASS**	**AVL Eventual Tar-**
System performance	3kW – 5kW	5kW	20 kW – 30 kW
Electrical efficiency	>50%	>50%	55% – 60%
Noise level	<60 dB	50dB	<40 dB
Start time	<30min	<15	<5 min
Weight/power	<20 kg/kW	-	3-4 kg/kW
Volume/power	<25l/kW	-	4-5l/kW
Lifetime	>500h	8000h	8000h

SOFC Range Extender (SOFC REX) System Design

The SOFC REX contains 4 x 28 cell SOFC stacks, which electrochemically convert, the partially reformed, ethanol fuel into electrical energy; and a Balance of Plant (BoP) that contains all the process for regulating the stack temperature, fuel supply and air supply. The MESTREX project is based around the initial development of a test bed system (REX 1) to validate the BoP operation and system controls, followed by a second system (REX 2), which will contain newly developed stack technology from Plansee SE and should meet the project technical targets. Concepts for both systems are shown in Figure 1. A key design feature of the REX 2 system is the < 60V stack Open Circuit Voltage (OCV). This is required in order to bypass high voltage safety requirements. Otherwise the system complexity would need to be significantly increased. Other key design considerations include; mechanical and electrical integration into the vehicle, a safety concept and a control strategy that allows a high fuel efficiency.

REX 1 AND REX 2 SYSTEMS

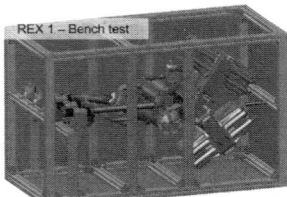

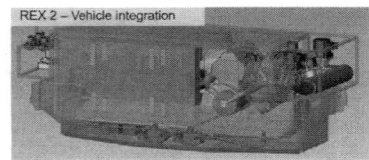

✓ Methods & Procedure for fast start and anode protection explored.
✓ Itterated BOP upgrades.
✓ Validate communication to vehicle.
✓ Test component performance
✓ Validate BOP side of fast start concept.
✓ Generate know how & experience to benifit Rex 2

✓ Similar basic control.
✓ Ready for vehicle installation.
✓ Start in 30 minutes.
✓ 6kW gross output.
✓ Ready for testing and evaluation in vehicle

Figure 1 (left) REX 1 bench test system (right) REX 2 for integration into a test carrier.

The process with the SOFC REX system is shown in Figure 2. The BoP contains an air blower, which is used to supply all the air requirements to the system. The SOFC stack, the start burner and the reformer. In the first systems developed in the MESTREX project, the air is supplied directly to the stack and is preheated by the hot gases exhausting the system in a heat exchanger (HEX). This air is used to pre-heat the stacks and regulate the stack temperature during operation. Since the SOFC stacks get hotter during operation, cooling is achieved via the stack cathode channel by supplying additional air. The endothermic steam reforming reactions in the anode channel are also used in order to support the stack cooling. Fuel is supplied to the system via injectors and supplied to the system fuel supply evaporator during hot operation. During start up the system start burner is also supplied with fuel via an injector.

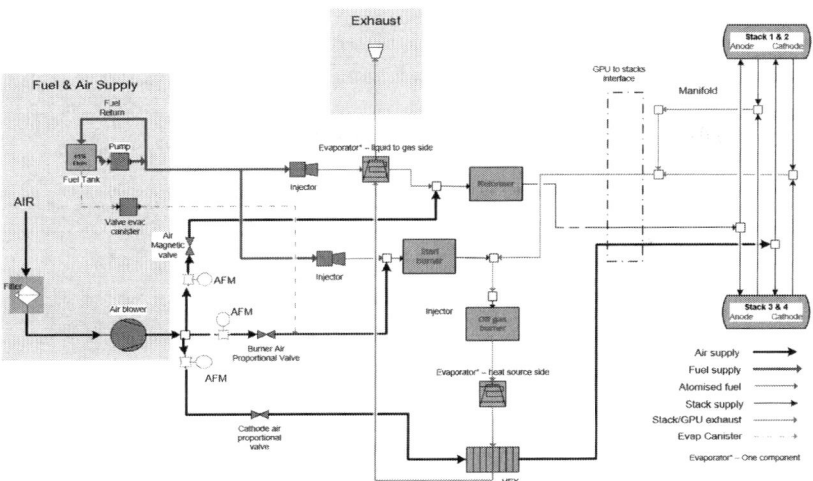

Figure 2: Flow sheet for the MESTREX system.

In MESTREX and COMPASS the fuel is a mixture of 45% ethanol and 55% water. The fuel is contained in the vehicle fuel tank. The evaporated fuel (and water) exhausts the evaporator and flows into the reformer unit, where steam reforming occurs. A reformer material screening task was executed within MESTREX in order to identify a supplier for the reformer catalyst coating. The reformers were tested with GHSVs that are in accordance with those expected within the MESTREX and COMPASS systems. Evidence of coking was also investigated and will require further investigations. In a next step, the selected coating will be evaluated with respect to sulfur tolerance. Since the reformer is exposed to an endothermic reaction, heat must be supplied externally. In Figure 3 the donut reformer and tube off gas burner are shown with some Computational Fluid Dynamics (CFD) results using the AVL-Fire™ software.

5

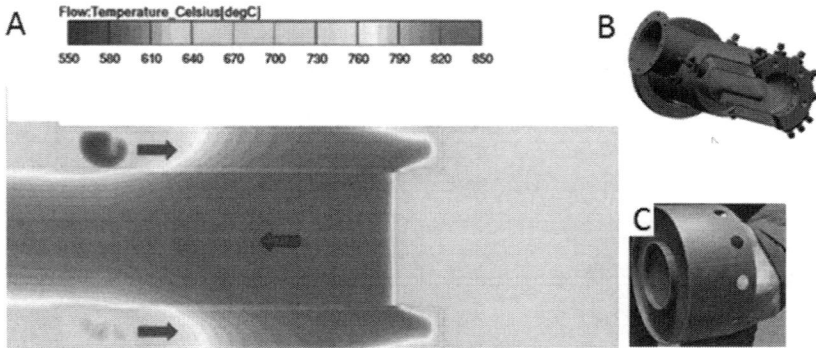

Figure 3: (A) Simulation results for the AVL reformer and off gas burner development (B/C).

A specialized burner test stand has been developed at AVL to characterize the combustion of the 45% ethanol 55% water fuel mixture. Stable and full combustion proved to be a challenge and several developments were required in order to find a solution. The test bench setup is shown in Figure 3. A combination of CFD and experimental activity were used to evaluate each burner design. For a normal flame combustor system, it is not possible to combust the fuel in a satisfactory manner and thus a catalytically active monolith with a preheating feature was selected, in order to light off and operate the start burner unit. With some geometrical and operating strategy optimizations, a start burner has been developed and integrated into the REX 1 system. Furthermore, on the REX 1 test stand several additional operating strategies can be trialed in combination with the off-gas burner and reformer to achieve the rapid system thermal activation and allow 5kW net electrical power within 30 minutes. Several methods that protect the anode on the REX 1 system, and if required on the REX 2 system, will also be trialed. These methods are related to and facilitated by the gas processing system and its design. In Figure 4 a solution was found for combusting the fuel mixture in a stable and satisfactory manner.

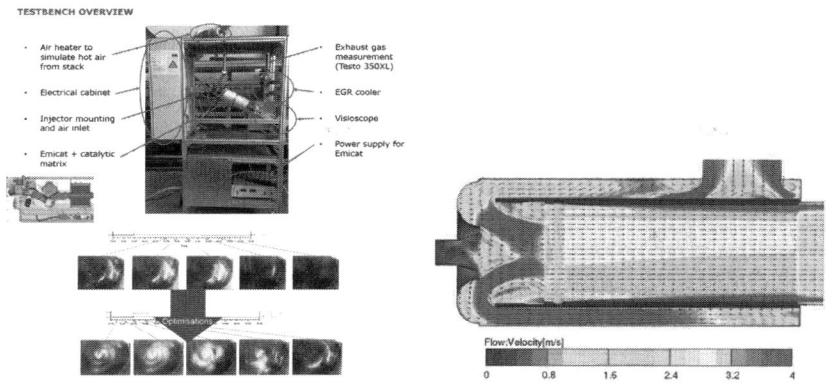

Figure 3. The burner test bench setup at AVL and optimization of an initial combustor design.

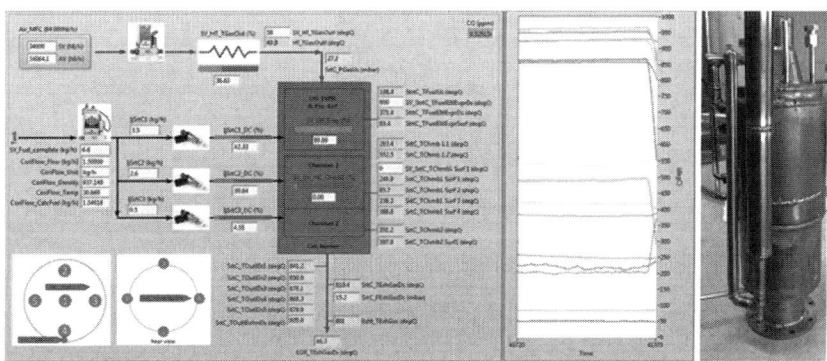

Figure 4: Burner development resultant of the efforts in MESTREX

Referring again to Figure 2, the fuel exhausts from the reformer and is distributed into the fuel (anode) channels within the SOFC stack. The stack operating temperature and tolerance to CH4, contained in the fuel, are two key parameters for designing the system BoP. Currently the stack operating temperature is expected to be in the range of 700 to 750°C in MESTREX from 650 to 700°C in COMPASS. AVL has designated a BoP simulation tool, which takes the key components, and component interactions (heat fluxes) shown in Figure 5 and outputs data which can be used in order to guide component and system design. For example, the variation of the stack temperature as a function of the operation of other components can be analyzed with AVL CAMEO™. This tool is used to design experiments and compare the actual performance of a system with the theoretical performance. When tuned for a system via a set of validation tasks, the software is used to pin point the components that can or should be optimized to reach a specific goal, such as; system efficiency, stack operating temperature or other criteria. Several features and operating strategies have been identified on the BOP side, which can make significant advances towards the > 50% efficiency target.

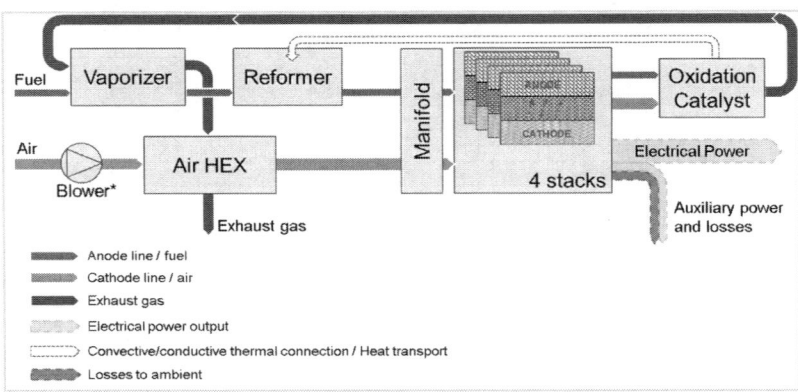

Figure 5: AVL designated simulation tool for SOFC BoP component & system, design, benchmarking and concept checking.

For the MESTREX REX 2 system AVL is also responsible for designing the gas manifold to the stacks. The stacks are provided in an open cathode format. Four stacks will be required in order to reach the 5kW net power requirement and an appropriate gas manifold and stack housing system has been designed. The challenge in designing this system was maintaining a homogeneous air and fuel mass flow through all four stacks, for a large variation in flow rates, without causing a significant pressure-drop or a complex manifold. AVL-fire ™ CFD has been used in for designing a manifold, and in a next step, a validation will be executed. A designated test bench has been designed to provide the required validation for the AVL-fire ™ simulation results.

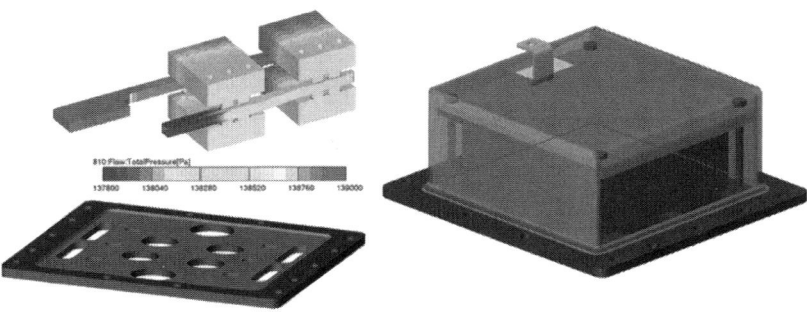

Figure 6: (Top Left) AVL-Fire ™ was used to design the stack manifolds (bottom and Right) AVL has designed a manifold and electrical isolation technology for mounting the stacks to the AVL BOP

The stacks to be used in MESTREX and COMPASS are a step beyond the state of the art SOFC technologies. They are based on metal substrates (Metal Supported Cells – MSC) and not ceramic (Electrolyte or Anode Supported Cells). The MSCs are expected to allow rapid starting and lower costs, but present some challenges including a gas tight design, corrosion tolerance and electrode, electrolyte and stack durability.

Figure 7: (Left) a Plansee stack for the SOFC rex (Right) example of single cells within the stack.

As referred to above, in COMPASS other options for rapidly heating the stack are being investigated, which will impose modifications on this flow sheet. The lower stack operating temperature will also mean some changes. COMPASS also builds upon MESTREX via the lessons learned within that project. AVL uses a Design Verification Validation Planning & Reporting methodology in order to generate a design and evaluation documentation and ensure that the system operation, durability and safety can be validated and verified within a systematic and focused documented database. In Figure 8 the fuel cell development in COMPASS is broken down into five levels and an example of validation and verification tests are shown in the table below. Each of these tasks is associated with a documented procedure and test result for establishing a record of the system during the development.

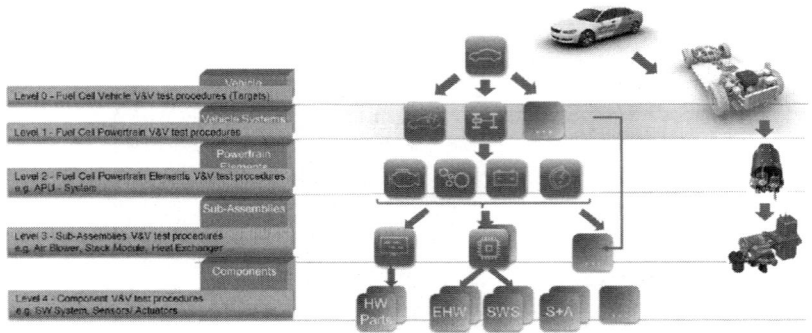

Figure 8: (TOP) Design validation and verification planning break down and (Bottom) Example of DVP documentation for the SOFC-REX system.

Conclusion

In the MESTREX project many of the challenges for operating a system on 45% ethanol and 55% water fuel have been identified and addressed. The system bench testing on the REX 1 rig are currently being executed and a rapid design of the REX 2 system has been planned for. The lessons learnt from the REX 1 system and component development can also be rapidly fed into the COMPASS project. In COMPASS, the focus is on reduced costs, rapid start and reduced weight and volume. Furthermore, in COMPASS Simulations with the vehicle using AVL CRUISE ™ are also being executed in order to investigate a vehicle operation with an SOFC-REX on board. COMPASS finally will realize the next step of the technology towards automotive application and the demonstration in a Next Generation Demo Car.

Acknowledgments

Part of this research was carried out in the project „ MESTREX" and is funded in part by the Austrian Ministry for Transport, Innovation and Technology (BMVIT) in the program "Mobility of the Future".

"Part of this research has received funding from the Fuel Cells and Hydrogen 2 Joint Undertaking under grant agreement No 700200. This Joint Undertaking receives support from the European Union's Horizon 2020 research and innovation programme and Hydrogen Europe and N.ERGHY."

Innovative propulsion systems with fuel cells

Assoc.Prof. DI Dr.techn. Manfred Klell,
HyCentA Research GmbH, CEO,
& Graz University of Technology

DI Dr.techn. Alexander Trattner,
HyCentA Research GmbH, CTO

© Springer Fachmedien Wiesbaden GmbH, ein Teil von Springer Nature 2018
J. Liebl (Hrsg.), *Der Antrieb von morgen 2018*, Proceedings,
https://doi.org/10.1007/978-3-658-21419-7_5

Introduction

The limitation of global warming to "less than 2 °C till the end of the 21st century compared to pre-industrial levels" as agreed upon during the 2015 Climate Change Conference in Paris can be regarded as the main challenge of our time [12]. As the anthropogenic climate change has far reaching consequences from weather extremes with increasing damage compensations, from health issues to population migration, the consistent decarbonisation of our economy is mandatory. Based on this insight, the UN and the EC have recently set targets for reduction of greenhouse gases on an international and a national level. E.g. in Austria a reduction of 36 % until 2030 (basis 2005) is required and a complete decarbonisation by 2050 is recommended in all relevant sectors, mobility and transport, households, industry and agriculture.

Electricity and hydrogen are the two only carbon-free energy carriers, and both of them have to be used in combination to enable the desired decarbonisation. So electricity has to be produced by wind, solar and waterpower, hydrogen can be produced via electrolysis as energy storage medium, which then can be stored, distributed and applied in traffic, industry and households; see Figure 1.

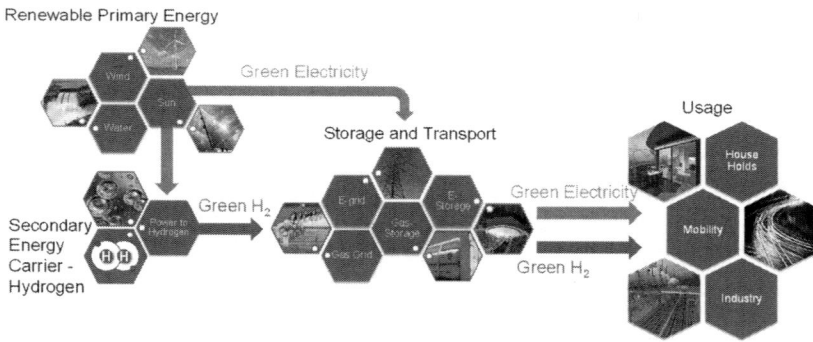

Figure 1: Green hydrogen in the energy and mobility system

Electrification of vehicles is an essential technological step in order to meet these challenges of the mobility sector. Moreover, mobility and transport technologies have to be completely transferred to zero-emission technologies in the long term to fulfil CO_2 fleet targets, to reduce pollutants especially in cities, and to enhance efficiency.

Zero Emission Vehicles

At present, battery electric vehicles (BEVs) feature high efficiency and low operation costs but the limited driving range (especially in cold conditions) and the long charging durations represent crucial drawbacks. The charging infrastructure for BEVs basically exists and the upgrade to higher charging power is ongoing. Nevertheless, a widespread installation of rapid charging, actual with a maximum power of 120 kW, will be limited by the costs for grid connection and the available and transferable amount of electric energy. Moreover, rapid charging may shorten the expected lifetime of batteries drastically.

Hydrogen-powered fuel cell electric vehicles (FCEVs) offer significant benefits. Hydrogen can be regarded as an unlimited, safe and efficient energy carrier produced from various sources which provides the required flexibility for future mobility. Basically, FCEVs provide similar driving ranges and refuelling durations (~3 minutes) compared to conventional ICE (Internal Combustion Engine) vehicles. Refuelling of 5 kg hydrogen in 3 minutes corresponds to a refuelling power of 3,4 MW. On account of the significant higher energy density of hydrogen compared to batteries [10,13], the sensitivity of the FCEV powertrain costs and weight to the amount of energy stored is low. Therefore hydrogen represents an appropriate fuel when high power and significant energy storage is required; see high weight and long range in Figure 2.

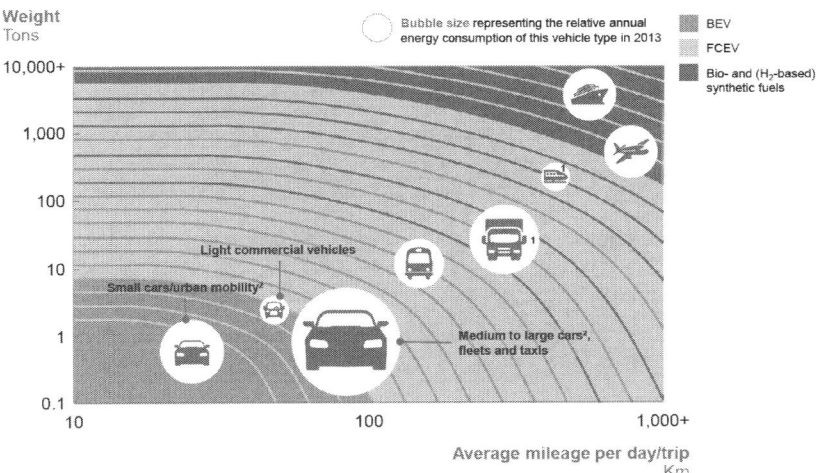

Figure 2: Role of FCEVs in decarbonizing transport [10]

BEVs and hydrogen-powered FCEVs represent zero-emission vehicles from Tank-to-Wheel (TtW). However, from Well-to-Tank (WtT) significant emissions can occur prior to the vehicle usage, depending on the type of electricity/hydrogen production and the charging/refuelling station concept [18]. Naturally, a significant increase of power generation from renewable energy sources as well as renewable hydrogen production by electrolysis is the prerequisite for low Well-to-Wheel (WtW) emissions of BEVs and FCEVs. The analyses of CO_2 emissions in Figure 3 are based on published data [4] of C-class vehicles: ICE – 5,4 l/100km fuel consumption, Diesel – 54 g_{CO2}/kWh WtT emissions; FCEV – 0,76 kg_{H2}/100km, SMR – 9,5 kg_{CO2}/ kg_{H2} WtT emissions, electrolysis with green electricity – 40 g_{CO2}/kWh; BEV – 20 kWh/100km, green electricity – 40 g_{CO2}/kWh. FCEVs powered by green hydrogen achieve very low WtW emissions. Even the usage of hydrogen produced by steam-methane-reforming (SMR) results in lower WtW emissions of FCEVs compared to those of conventional ICE-powered Diesel vehicles. BEVs powered by green electricity achieve lowest WtW CO_2 emissions.

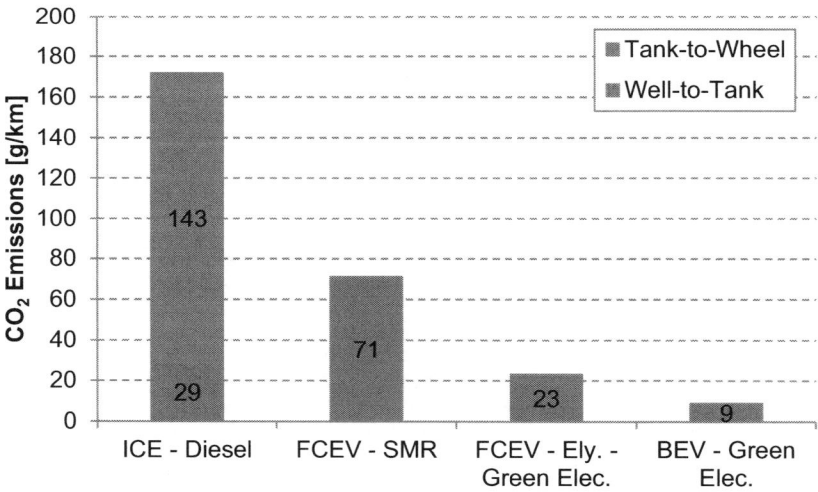

Figure 3: CO_2 emissions of different propulsion systems

For the reduction of overall CO_2 emissions the overall life cycle including production and recycling needs to be considered in addition to the WtW emissions. Life-cycle-analyses (LCAs) of FCEVs are based on limited data as only few are in series production. Nevertheless, first LCAs [10,19] indicate very low CO_2 emissions of the overall life cycle of FCEVs with the result that FCEVs powered by green hydrogen achieve lowest emissions of all powertrains. BEVs have higher life cycle emissions as production and recycling of batteries is very energy and resource intensive.

FCEV Powertrain Concepts

The powertrain of FCEVs basically consists of hydrogen storage system (energy storage), fuel cell system (energy converter), battery system (energy storage and converter), various voltage transformers (converters and inverters), e-motor, gear box and mechanical drive to the wheels. Hence, FCEVs represent electro-hydrogen-hybrids whereas batteries and fuel cells ideally complement each other. Basically, FCEVs are classified into two main concepts: the fuel cell dominant concept and the range extender (REX) concept. Moreover, combinations of these are possible, which are usually mentioned as "mid-size fuel cell" concepts.

At the fuel cell dominant powertrain concept the required driving performance is covered by the fuel cell, the battery is used to recuperate braking energy and to provide additional energy during acceleration. Hence, for automotive applications the fuel cell power typically ranges between 100 to 150 kW, the battery features 1 to 2 kWh with high power (high C-rates) and the hydrogen storage features 5 to 6 kg of hydrogen to reach driving ranges of approximately 600 km. Energy supply to the vehicle is solely performed by refuelling with hydrogen.

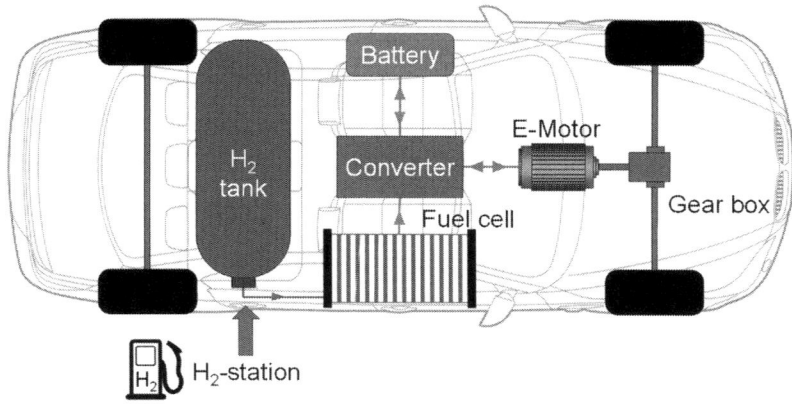

Figure 4: Fuel cell dominant powertrain concept [13]

At the REX powertrain concept the required driving performance is covered by the battery, the fuel cell charges the battery during driving when state-of-charge (SOC) falls below a certain level in order to extend the driving range. REX vehicles usually feature batteries with high capacity (>10 kWh), fuel cell power of 20-30 kW, and smaller hydrogen storages compared to the fuel cell dominant concept. Energy supply to the vehicle is performed by charging the battery (plug-in) and refuelling with hydrogen.

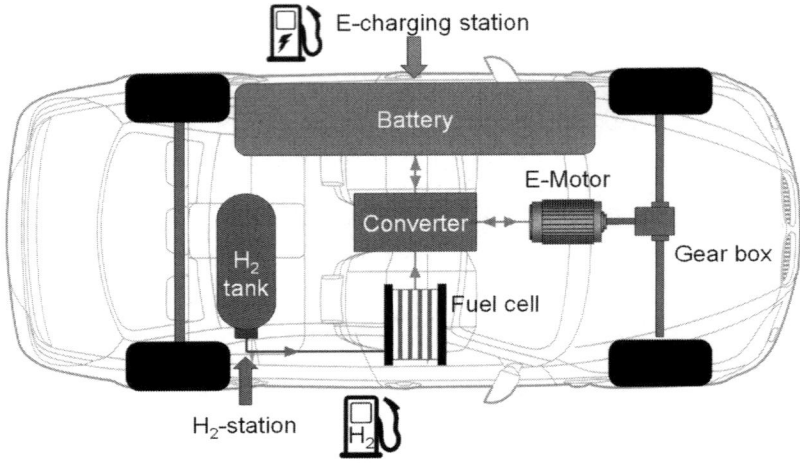

Figure 5: Fuel cell range extender concept [13]

The present high costs of FCEVs at low production volume represent the main barrier for the widespread introduction. At high production volume the FCEV achieves lower costs than BEVs when high driving range is required, see Table 1. This has been calculated based on the published cost numbers for high production volumes of 500.000 #/a [3,6]: battery costs of 112,5 €/kWh, fuel cell costs of 40 €/kW and hydrogen storage costs of 300 €/kg$_{H2}$. Boundaries of the calculated vehicle are: vehicle energy consumption of 15 kWh/100km, energy utilization of battery of 80 %, TtW efficiency of BEV of 84 %, and hydrogen consumption of FCEV of 0,85 kg/100km.

Table 1: Cost comparison of FCEV and BEV at high volume production

	Driving Range in km	100	200	300	400	500	600
FCEV	PEMFC - 100 kW	€ 4.000	€ 4.000	€ 4.000	€ 4.000	€ 4.000	€ 4.000
	Battery 2 kWh	€ 225	€ 225	€ 225	€ 225	€ 225	€ 225
	Hydrogen Storage System	€ 255	€ 510	€ 765	€ 1.020	€ 1.275	€ 1.530
BEV	Energy net in kWh	15	30	45	60	75	90
	Energy name-plate in kWh	22	45	67	90	112	134
	Costs at € 112,5 per kWh	€ 2.517	€ 5.035	€ 7.552	€ 10.070	€ 12.587	€ 15.105
Vehicle	FCEV	€ 4.480	€ 4.735	€ 4.990	€ 5.245	€ 5.500	€ 5.755
	BEV	€ 2.517	€ 5.035	€ 7.552	€ 10.070	€ 12.587	€ 15.105

PEM Fuel Cell

The fuel cell as an electrochemical energy converter is able to directly convert the inner chemical energy of the fuel, usually hydrogen, into electrical energy. Hydrogen powered PEMFCs (proton exchange membrane fuel cells) are actually preferred for mobile applications as they offer a variety of advantages like high efficiency, high volumetric as well as gravimetric power density and the potential to achieve low costs in mass production. The electrodes typically consist of Pt-based catalysts separated by proton conducting perfluorosulfonic acid (PFSA) or hydrocarbon membranes, whereby membrane thicknesses below 25 μm are achieved. As the voltage of a single fuel cell is limited, a number of cells are combined to form a fuel cell stack. This poses several challenges, since pressures, temperatures, and mass flows of supply media have to be controlled within tight limits. In order to operate the fuel cell stack properly, it is necessary to control a number of auxiliary components which form the fuel cell system including energy and thermal management [16]. Within the last years significant technical progress has been achieved in terms of platinum content reduction, reduction of degradation, and improvement of freeze start behaviour; see Figure 6. Hence, scale-up and industrialization has been started. The general objective in the FC development is to significantly reduce the costs and system degradation in order to increase the market penetration of FC vehicles. Actually, the main research and development tasks are more and more focussing on system level, production technologies and quality control. Actual cost levels predicted to automotive relevant production volumes are in the range of 50 US$/kW, the long term target of the DOE is 30 US$/kW; see Figure 6.

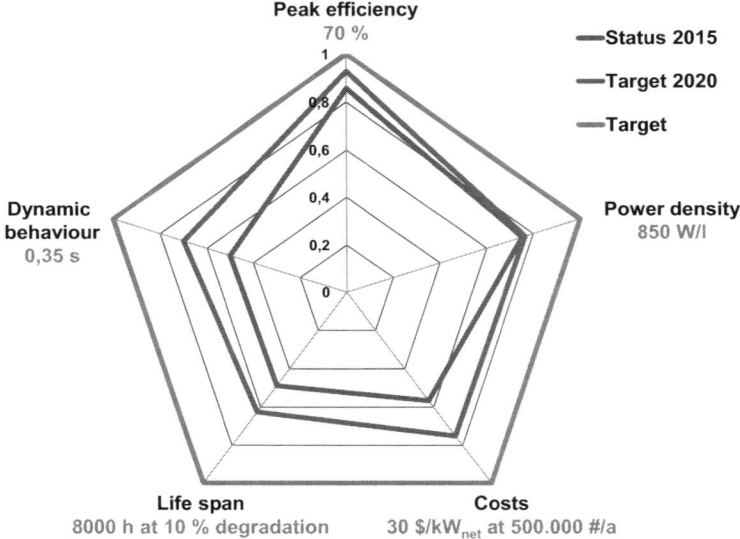

Figure 6: Status of automotive PEM fuel cell systems [3]

In order to perform applied research in the area of PEMFC systems, a highly integrated fuel cell analysis infrastructure for systems up to 160 kW electric power has been developed and established within the cooperative research project HIFAI-RSA by Hy-CentA Research GmbH and AVL List GmbH in Graz, Austria, see Figure 7.

Figure 7: HIFAI-RSA: Highly Integrated Fuel Cell Analysis Infrastructure [2]

Vehicle, driver and driving cycle as well as powertrain components like battery, electric engine, transmission and different balance of plant (BoP) components can be simulated in real time with hardware in the loop (HiL), the fuel cell system is physically tested. Ambient conditions and media supply temperatures can be adjusted dynamically in the range of −40 °C to 85 °C. Moreover, cathode air humidity can be varied in the range of 5 % to 95 %. The test bed enables research and development on topics from energy management to thermal management, from complete vehicle to sub-system control and calibration, from vehicle integration to the investigation of dynamics, cold start and degradation. Moreover, calibration and optimization can be performed automatically based on DOE tools.

Hydrogen Storage Systems

To date, FCEVs are fuelled with gaseous hydrogen at pressures of 35 MPa to 70 MPa. As 70 MPa tanks allow for much higher ranges at acceptable tank volumes, most recent vehicles are equipped with these. Even high energy density batteries like Li-Ion batteries, achieve only a gravimetric energy density an order of magnitude lower than compressed hydrogen; see Figure 8.

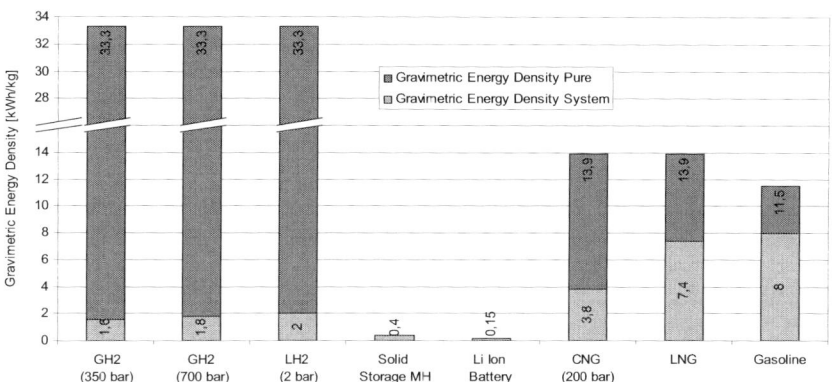

Figure 8: Gravimetric energy density of energy storage systems [13]

The fuelling process is regulated in SAE J2601 including fuelling time and related pre-cooling procedures. Basically, the packaging of the hydrogen storage system in the vehicle represents a major challenge due to the cylindrical shape of the vessels and the lower volumetric energy density of gaseous hydrogen compared to fossil fuels. Actual development is focusing on simplification of the overall tank system including functional integration of single components in order to further reduce costs.

Fuel Cell Applications

Fuel cell propulsion systems are seen as the "long distance" and "all purpose" alternative to existing pure battery electric propulsion systems. Actually fuel cells are applied in several applications, either in air, on sea or on land; see Figure 9. Industrialisation has been started, fuel cell market growth is accelerating and the transfer to additional applications is ongoing.

Figure 9: Overview of fuel cell applications [13]

In Austria the energy model region WIVA P&G (Wasserstoffinitiative Vorzeigeregion Austria Power & Gas) with a project volume of approximatelly 120 M€ illustrates the conversion of the Austrian energy system to green hydrogen and green electricity funded by the Austrian government and the climate and energy fund. Several mobility applications (special vehicles, car fleets, busses, trucks etc.) will be developed and demonstrated in real usage. In the following, ongoing research, development and demonstration projects in Austria, and status of technology of FCEVs are presented.

Special Vehicles

PEMFC technologies can be used in a whole range of special vehicles used for various specific purposes like utility vehicles (construction, agriculture, forestry, logistics, etc.) and leisure vehicles (sport and tourism). Today most of these vehicles are still powered by ICE whereby the electrification of these offers a great reduction potential for GHG emissions and the chance to enter new business fields. In the course of the HySnow project, funded by the Austrian climate and energy fund, off-road vehicles for winter tourism with PEMFC propulsion system and a de-centralized green hydrogen infra-structure bases on high-pressure PEM electrolysis is developed. A pure battery-electric powertrain can be excluded a priori as batteries show several disadvantages in cold weather conditions like low energy capacity, leading to low driving range and bad re-charging behaviour. The most promising zero-emission powertrain technology for off road vehicles for cold conditions is the hydrogen-driven PEMFC. The advantages are: operability in cold conditions, simplicity, high efficiency, quick start-up, modularity, and high power density. A hydrogen-based PEMFC provides similar power, range and refuelling time as a modern gasoline powered vehicle. As the efficiency of the fuel cell is considerably higher than that of the ICEs, more than double, the fuel cell is an ideal technology to convert chemically stored energy in hydrogen into electricity on-board and, thus, solve the range and charging issues of battery EVs. Innovations include the solving of the following challenges: functionality of the whole chain in the real-life operating conditions of freezing winter temperatures, freeze starts of the fuel cell, min-imisation of the degradation, driveability of the vehicle, efficiency of the powertrain, the vehicle, and the infrastructure, durability of the powertrain, the vehicle, and the infrastructure. Special emphasis will be put from the beginning in the reduction of costs of the whole chain. Product and component design, chosen materials, and optimised operation strategies will be evaluated to minimise costs.

Passenger Cars

Since 2014 several Asian car manufacturers have introduced series production FCEVs (Hyundai ix35 in Figure 10, Toyota Mirai in Figure 11, Honda Clarity in Figure 12) demonstrating technological feasibility and outstanding advantages for the customers. These FCEVs feature high efficiencies as well as low hydrogen consumption (<1 kg/100 km), low noise, excellent driveability, high range, short refuelling duration (max. 3 min) and zero-emissions. All these actual series production FCEVs are hybrids consisting of a strong fuel cell with a peak power of approximately 100 kW and a small battery (NiMH or Li-ion) with 1-2 kWh capacity and up to 20 kW power. This config-uration increases efficiency, enables regenerative braking and further improves dy-namic response of the vehicle. In 2018 Mercedes starts with a small series production of the Mercedes GLC F-Cell, see Figure 13. The GLC F-Cell is a plug-in hybrid with a driving range of 437 km based on the stored 4,4 kg hydrogen. The vehicle achieves a maximum power of 147 kW and is equipped with a 13,8 kWh battery.

11

Figure 10: Hyundai ix35 FCEV [11]

Figure 11: Toyota Mirai [20]

Figure 12: Honda Clarity [8]

Figure 13: Mercedes GLC F-Cell [14]

The transnational project KEYTECH4EV, a consortium under the lead of AVL, follows a more innovative hybridization approach and is currently developing a demonstration fuel cell car in their flagship project KEYTECH4EV. The project KEYTECH4EV consists of an industrial engineering provider (AVL List GmbH), component and subsystem manufacturers (ElringKlinger AG, Hörbiger Ventilwerke GmbH, Magna Steyr AG), research institutes (Graz University of Technology, HyCentA Research GmbH, Vienna University of Technology) and the small and medium enterprise IESTA. The project KEYTEC4EV develops key technologies for the integration and build-up of a hybridized demonstration vehicle with a cost-optimized and CO_2-free powertrain based on fuel cell and battery technology. Intensive cost analyses based on a Toyota Mirai deep-dive benchmark indicate for the timeframe 2020-2030 that a fuel cell – battery hybrid will reach lower cost than a pure battery EV or a pure fuel cell EV (with very small battery). The key technologies comprise the stack, fuel cell system, injector/ejector anode system, hydrogen-tanks, thermal management, energy management, and controls.

Figure 14: Vehicle layout – KEYTEC4EV

The VW Passat GTE was chosen as the vehicle platform for all development activities, see Figure 14. The pre-existing battery system (battery pack, battery cooling and plug-in charging including power electronics) will be utilized, whereas all other powertrain components will be replaced by KEYTECH4EV technology to create an innovative electrified powertrain (e-motor, power electronics, fuel cell system, control hardware and software); see technical data in Table 2. As such, this conceptual approach will increase public acceptance of electro-mobility as it provides driving distances of above 600 km per fuelling and further benefits of short refilling times of about 3 minutes. At the same time, the overall powertrain costs are reduced in comparison to FC dominant powertrain concepts and pure BEVs.

Table 2: Technical Data of KEYTEC4EV

Vehicle platform	VW Passat GTE
Battery capacity / power	9.9 kWh / 85 kW
Fuel cell system power	~60 kW
e-drive power	110 kW
Hydrogen tank capacity	5.3 kg
Hydrogen consumption	0.85 kg /100 km
Driving range	>600 km (NEDC, 5 kg H_2 incl. battery-only range)

Vans

In the framework of the funded research project FCREEV [7,17], the consortium formed by Magna Steyr Engineering AG & Co KG, Institute for Powertrains and Automotive Technology (Technical University of Vienna), HyCentA Research GmbH and Proton Motor Fuel Cell GmbH has investigated and developed a fuel cell range extender concept and created a drivable technology; see Figure 15. Powertrain costs of BEVs and FCEVs significantly depend on production quantity. Concerning the costs at low quantities, the fuel cell range extender concept has the potential to lower costs and even fall below the costs of BEVs by exceeding 100-150 km driving range.

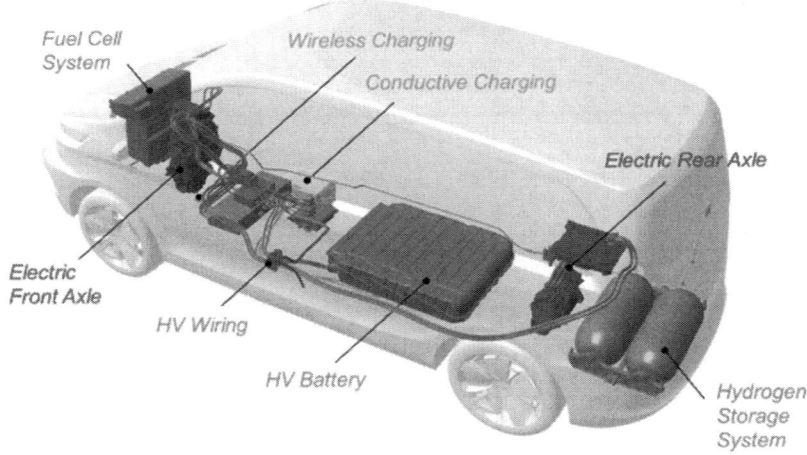

Figure 15: FCREEV [17]

The electric powertrain consist of the electric drive systems (consisting of DC/AC inverter, electric motor and single-stage transmission) for the front and rear axle of the vehicle, the fuel cell system (FC stack, air compressor, DC/DC inverter and control unit), the HV battery (Li-ion) as well as different charging systems and power electronics for voltage conversion in the power supply system.

Pure battery electric driving is allowed at a high SOC (state of charge) to force discharge of the battery. With lower SOC levels a sophisticated operating strategy activates the fuel cell to achieve a maximum range and highest efficiency (54 %) of the 25 kW fuel cell system. With this configuration driving ranges up to 70 km are possible without support of the fuel cell thus achieving the average daily range of customers. Range extenders, however, are not primarily used for supplying power at any operational level but for covering the average power demand. Peak power demands and regenerative braking energy are covered by charging and discharging of the battery. The combination of both energy sources ensures higher power demands and longer driving ranges (> 350 km) without refuelling or recharging. Due to the FCREEV concept the fuel cell is frequently operated in stationary and low degradation conditions. Test bench investigations resulted in degradation less than 7 % after 5000 operating hours to ensure the power supply by the fuel cell system for the complete lifetime.

Heavy-Duty Applications

Considering the high energy density of hydrogen and the advantages of FCEV powertrains, FCEVs will be especially important in decarbonizing buses, heavy-duty transportation, and non-electrified trains. The application of PEMFCs in the heavy duty sector is considerable increasing and has been completely underrated up to now.

Figure 16:
FC bus – Toyota [20]

Figure 17:
FC truck – COOP [9]

Figure 18:
FC train – Alstom [1]

Fuel cells busses have been thoroughly proven since years and several hundred busses are in operation world-wide. In Europe the FCH-JU is funding bus demonstration projects whereas approximately 84 busses are in operation in Europe [5]. Especially in Asia R&D and demonstration activities are rapidly increasing; e.g. Lianyungang Haitong Public Transport (China) plans for 1500 FCEV buses and Toyota plans to introduce over

100 FC buses, ahead of the Tokyo 2020 Olympic and Paralympic Games; see Figure 16. Basically, fuel cell systems for busses are designed to achieve low degradation whereas degradation rates of 10 % after 10.000 h have been demonstrated in the past [15]. The development of the next generation of fuel cell systems for heavy duty applications aims for 25.000 h. Moreover, fuel cell busses with a power of 200 kW are consuming 8 to 9 kg of hydrogen and achieve driving ranges of 300 to 450 km [5,15]. Hence, FC busses offer high durability, high TtW efficiency, and zero-emissions with the flexibility of a Diesel bus. The same applies for FC truck applications. Several truck concepts are in development (Toyota, Nikola Motor Company, and COOP in Figure 17) in order to fulfil the need for zero-emission logistics in cities.

For non-electrified lines where mainly Diesel ICE trains are operated today, the PEMFC powertrain offers a zero-emission alternative. Germany announced recently that its first hydrogen trains are running. Alstom has developed the Coradia iLint which features a power of 400 kW and a driving range of 600 to 800 km based on the stored 180 kg of hydrogen; see Figure 18. Moreover, FCEV trains are already cost competitive with diesel trains (from a TCO perspective).

Summary

The major challenges for future mobile and road transport and therefore the global automotive industry are enabling individual mobility, reduction of the energy consumption and decarbonisation. To meet these challenges, electric vehicles with battery and fuel cell technology will be the key enabler. Pure battery electric vehicles are suitable for smaller vehicles and short ranges (urban) whereas fuel cells are more appropriate for larger vehicles with longer ranges and higher power demand.

At FCEVs energy conversion and energy storage are separated, the FC is the energy converter and the hydrogen tank the energy storage, which allows for a higher energy stored on-board and therefore higher driving range and faster refuelling compared to BEVs. FC propulsion systems are seen as the "long distance" and "all purpose" alternative to existing pure battery electric propulsion systems. FCEVs will play a major role in decarbonizing mobility as FCEVs powered by green hydrogen achieve lowest emissions of all powertrains over the whole life cycle. Moreover, FCEVs represent a cost-effective solution as at high production volume the FCEV achieves lower costs than BEVs when high driving range is required. The FCEV technology is perfectly suited for the development in Europe as existing know-how, manufacturing technologies and supply chains can be transferred from ICE development and production.

Actually FCs are applied in several applications, either in air, on sea or on land. In the course of the energy model region WIVA P&G (Wasserstoffinitiative Vorzeigeregion Austria Power & Gas) the conversion of the Austrian energy system to green hydrogen and green electricity is demonstrated. In several Austrian research, development and demonstration projects various vehicle concepts like snow groomers, passenger cars, vans, busses, trucks, and trains are investigated. Finally, hydrogen's unique properties like long term and efficient energy storage make it a powerful enabler for the energy transition, with benefits for the energy system and end-use applications.

References

1. Alstom, http://www.alstom.com/press-centre/2017/03/alstoms-hydrogen-train-coradia-ilint-first-successful-run-at-80-kmh/ , (20.10.2017)
2. Brandstätter, S., Striednig, M., Aldrian, D., Trattner, A. et al., "Highly Integrated Fuel Cell Analysis Infrastructure for Advanced Research Topics," SAE Technical Paper 2017-01-1180, 2017, https://doi.org/10.4271/2017-01-1180.
3. DOE – Department of Energy USA, DOE Technical Targets for Fuel Cell Systems and Stacks for Transportation Applications, https://energy.gov/eere/fuelcells/doe-technical-targets-fuel-cell-systems-and-stacks-transportation-applications (20.10.2017)
4. Edwards, R., Larive´, J.-F. , Rickeard, D. and Weindorf, W., "Well-to-Wheel Analysis of future Automotive Fuels and Powertrains in the European Context.", JEC Technical Reports, 2014, doi:10.2790/95629.
5. FCH JU: Fuel Cell Electric Buses – Potential for Sustainable Public Transport in Europe, The Fuel Cells and Hydrogen Joint Undertaking, 2015
6. Grögera, O., Gasteiger, H., Suchsland, J.: Review—Electromobility: Batteries or Fuel Cells?, Journal of the Electrochemical Society, http://jes.ecsdl.org/content/162/14/A2605.full , 2015.
7. Höflinger, J.; Hofmann, P.; Müller, H.; Limbrunner, M.: FCREEV – A Fuel Cell Range Extended Electric Vehicle. In: MTZ ww 77 (2017), No. 5, pp. 16-21
8. Honda Motor, http://world.honda.com/FuelCell; http://www.honda.de/cars/honda-welt/news-events/2015-10-28-honda-enthuellt-clarity-fuel-cell-auf-der-tokyo-motor.html (24.07.2017)
9. H2energy, http://h2energy.ch/wp-content/uploads/2017/06/Factsheet_Lastwagen_D.pdf , (10.12.2017)
10. Hydrogen Council, http://hydrogencouncil.com/ , (20.10.2017)
11. Hyundai Motor Company, https://www.hyundai.com/worldwide/en/eco/ix35-fuelcell/highlights
12. IPCC: Climate Change 2014: Synthesis Report. Contribution of Working Groups I, II and III to the Fifth Assessment Report of the Intergovernmental Panel on Climate Change. IPCC, Geneva, Switzerland, 151 pp, 2014

13. M. Klell, H. Eichlseder, A. Trattner: Wasserstoff in der Fahrzeugtechnik, Springer Verlag, Buch 4. Auflage, 2018 (in Vorbereitung)

14. Mercedes-Benz, https://www.mercedes-benz.com/de/mercedes-benz/fahrzeuge/personenwagen/glc/der-neue-glc-f-cell/ (20.10.2017)

15. Müller, K.; Schnitzeler, F.; Lozanovski, A.; Skiker, S.; Ojakovoh, M.: Clean Hydrogen in European Cities, D 5.3 – CHIC Final Report, FCH JU, 2017

16. Nöst, M.; Doppler, Ch.; Klell, M.; Trattner, A.: Thermal Management of PEM Fuel Cells in Electric Vehicles, Buchkapitel, Comprehensive Energy Management, Seite 93-112, Springer, ISBN 978-3-319-57444-8, 2017

17. Salman, P., Wallnöfer-Ogris, E., Sartory, M., Trattner, A. et al., "Hydrogen-Powered Fuel Cell Range Extender Vehicle – Long Driving Range with Zero-Emissions," SAE Technical Paper 2017-01-1185, 2017, https://doi.org/10.4271/2017-01-1185.

18. Sartory, M., Justl, M., Salman, P., Trattner, A. et al., "Modular Concept of a Cost-Effective and Efficient On-Site Hydrogen Production Solution," SAE Technical Paper 2017-01-1287, 2017, https://doi.org/10.4271/2017-01-1287.

19. Tokieda, Junji: „The Mirai – Life Cycle Assessment Report", Toyota, 2015

20. Toyota Motor Corporation, http://www.toyota.com, https://www.toyota.at/new-cars/new-mirai/index.json#1 , (20.10. 2017)

A comparative evaluation on state-of-charge estimation methods for Lithium-ion batteries of electric vehicles

Fuliang HUANG, Masashi MUROHOSHI, Akira ICHINOSE, Tingting SUI

Keihin Corporation, 2021-8, Hoshakuji, Takanezawa-machi, Shioya-gun, Tochigi, 329-1233, Japan

© Springer Fachmedien Wiesbaden GmbH, ein Teil von Springer Nature 2018
J. Liebl (Hrsg.), *Der Antrieb von morgen 2018*, Proceedings,
https://doi.org/10.1007/978-3-658-21419-7_6

Abstract

Due to strengthening enforcement of environmental laws and regulations, electrification of the automotive powertrain has been advancing so rapidly that rechargeable batteries used as a source of power are becoming increasingly important. Among them, lithium-ion batteries (LiBs) are expected to evolve at a high rate regarding light weight, high energy density and long lifetime. In order to make the most capacity of lithium-ion batteries in the strict automotive environment, the battery management system (BMS) constantly monitors the state of the batteries, and estimates the state-of- charge (SOC) for avoiding overcharge / overdischarge of the batteries. This paper focuses on the existing methods for estimating SOC from voltage, current and temperature of automotive LiBs, performs comparative evaluation, and proposes a process for SOC estimation.

Keywords: BMS, LiB, SOC estimation, OCV, EKF

1 Introduction

The spread of motorized automobiles such as electric vehicles (EVs) has been increasing worldwide. France and the UK have announced a policy to ban the sale of vehicles powered by only internal combustion engines from 2040. Subsequently, India has also launched a policy to sell only EVs by 2030. China announced that, New Energy Vehicles (NEVs) should reach at least 10% of the sales quantity in 2019 – targeting 7 million NEV sales annually by 2025, and expressed that the timing to completely prohibit gasoline vehicles in the future is under review. However, the spread of EVs is faced with various challenges. LiBs with high energy density and high charge and discharge efficiency are commonly used in battery powered EVs, but, it is necessary to install far more batteries to attain a driving distance which is comparable to current gasoline vehicles. Furthermore, it is difficult to overcome the situation of battery degradation during driving [1]. Driving without understanding the situation of battery degradation may cause unforeseen events. Thus, for safely and reliably using power battery in EVs, it is necessary for the BMS to accurately determine the capacity of the battery at any one time using a high performance battery electronic control unit (ECU) and high precision sensors [2][3].

Currently, LiBs used in EVs are utilizing different materials (e.g. cathode material containing NCA, NMC, LMO, LFP, etc.) and cell formats (e.g., cylindrical, prismatic or pouch cell), resulting in different electrochemical and electrophysical properties. An approach to LiB capacity estimation, typically represented by SOC, generally utilizes the open circuit voltage (OCV) method, the current integration method, modeling method using a Kalman filter as an example, or their combination.

As known, the OCV method based on the pre-assumption that there exists an approximate linear relationship between the OCV and the SOC, is most widely used but has drawbacks in that, it is invalid during battery operation, and polarization relaxation requires long time (2 hours after disconnected from the load). Therefore, OCV tables and maps prepared beforehand are required. Another method widely used is the current integration method which integrates the charging and discharging currents for estimating the SOC. While the charging and discharging current is accumulated continuously over a long period, the accumulated initial error and measurement error also gradually increase and this causes the estimated SOC to diverge from the actual SOC. Therefore, correction is required to minimize the error. The modeling method, for instance, using an equivalent electrical circuit model to investigate electrochemical impedance (including adaptive digital filtering algorithm like a Kalman filter), has trade-off relationship between the accuracy of the model and the accuracy of state estimation. An increment of model parameters improves the accuracy of the model, but impairs the accuracy of state estimation. That is to say, the setting parameters is complicated and difficult in practical use.

This paper focuses on the derivation of algorithms for high accuracy SOC estimation, from the existing OCV, current integration, and modeling methods. The two types of cylindrical LiB (i.e., 18650 NCA, and 18650 NMC) that are most used for EVs (hereafter referred to as LiB-A and LiB-B separately) are introduced and pretested in Chapter 2. Chapter 3 describes the model construction for LiB-A. Chapter 4 provides simulation results for existing algorithms on LiB-A, and discusses improvement in using these algorithms through comparative analyses. Subsequently, improvements are applied to LiB-B in Chapter 5 to demonstrate the effectiveness. Chapter 6 briefly summarizes this paper, and suggests the direction for further method improvement.

3

2 Battery Specifications

Table 1 shows specifications of the battery cells, LiB-A and LiB-B, to be used in the comparative evaluation.

Table 1 Specifications of LiB-A and LiB-B

	LiB-A	LiB-B
Type	18650 NCA	18650 NMC
Nominal Voltage (V)	3.6	3.6
Charging Voltage (V)	4.2	4.2
Discharging Voltage (V)	2.5	2.5
Capacity (mAh)	3250	3000
Temperature		
Charge (℃)	0 to 45	0 to 50
Discharge (℃)	-20 to 60	-20 to 75
Wight (g)	48.5	48
Energy density		
Volumetric (Wh/l)	676	661
Gravimetric (Wh/kg)	243	240

Source: data released on battery manufacturer's official website

In order to accurately ascertain the battery cell charge and discharge characteristics, a test system configured by the KIKUSUI charge and discharge system controller PFX2512, in combination with the DC power supply PWR400L and electronic load PLZ164W, shown in Figure 1, has been introduced [4]. The exclusive application software BPChecker3000, supports seamless charge and discharge and high speed data sampling. The main features of this system are: a maximum voltage of 60.0000V; a maximum current of 50.0000A; measurement precision of capacity, voltage and current of 0.1mAh, 0.1mV and 0.1mA respecttively; a maximum sampling speed of 1ms; pattern charging and discharging capabilities by 10000 steps; and temperature measurement during charging and discharging, thus ensuring accurate testing.

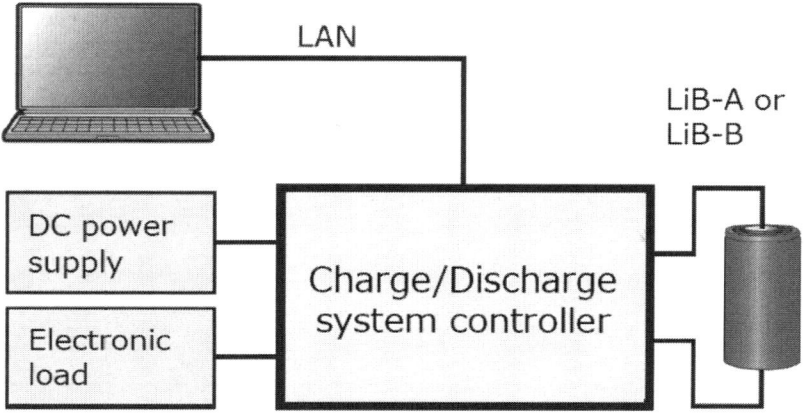

(Source: KIKUCHI Charge/Discharge System Controller - System configuration)

Fig.1 Test environment for LiB cells

The battery cell charge and discharge tests were carried out at three temperatures of 0□, 25□, and 45□ to confirm OCV-SOC and I-V characteristics by using new battery cells.

Charging procedure:

1. Set temperature of constant temperature chamber, perform constant current (CC) charging at 1C rate from the lower limit of cell reference capacity. When voltage reaches 4.2V, terminate the CC charging.
2. Then, perform constant voltage (CV) charging. Terminate the CV charging when charging current drops to 1/50C rate or less.
- SOC at this point of time is considered to have reached full charge capacity (FCC) status, and was defined as 100%.

Discharging procedure:

Set temperature of constant temperature chamber, repeat discharging step of, 3 minutes' CC discharge and 60 minutes pause, at 1C rate. Terminate discharging when terminal voltage of battery drops to 2.5V.

- The terminal voltage after the 60 minutes pause is regarded as open circuit voltage.

5

Figure 2 a), b) and c) show the charge characteristics of LiB-A at 0□, 25□, and 45□; Figure 2 d), e) and f) show the corresponding discharge characteristics of LiB-A at 0□, 25□, and 45□, through pretest respectively. Figure 3 shows charge and discharge characteristics of LiB-B at 0□, 25□, and 45□, comparatively.

The measured capacity of LiB-A, using cathode materials of lithium nickel cobalt aluminum oxide (NCA), wss 3.08Ah, compared with the nominal capacity of 3.25Ah, and the measured capacity of LiB-B, using cathode materials of lithium nickel cobalt manganese oxide (NMC), was 2.82Ah, compared with the nominal capacity of 3.00Ah. That is to say, LiB-A and LiB-B are qualified for performing comparative evaluation of SOC estimation methods.

Figure 4 a), b) and c) show OCV-SOC lookup tables for LiB-A at 0□, 25□, and 45□; Figure 4 d), e) and f) show impedance-SOC lookup tables for LiB-A at 0□, 25□, and 45□ respectively. Figure 5 shows OCV-SOC and impedance-SOC lookup tables for LiB-B at 0□, 25□, and 45□, respectively.

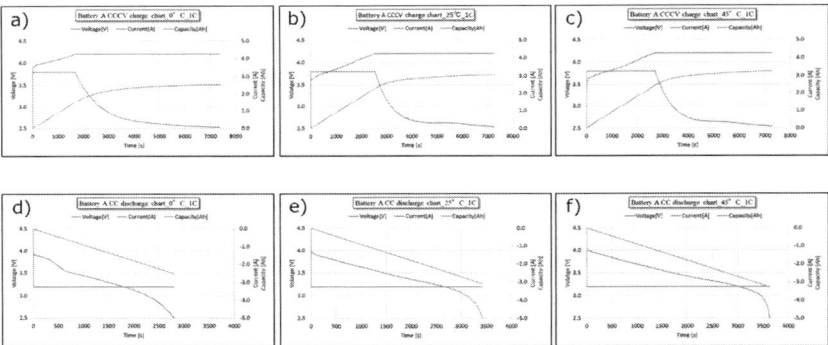

Fig.2 Charge and discharge characteristic curves of LiB-A at 0□, 25□ and 45□

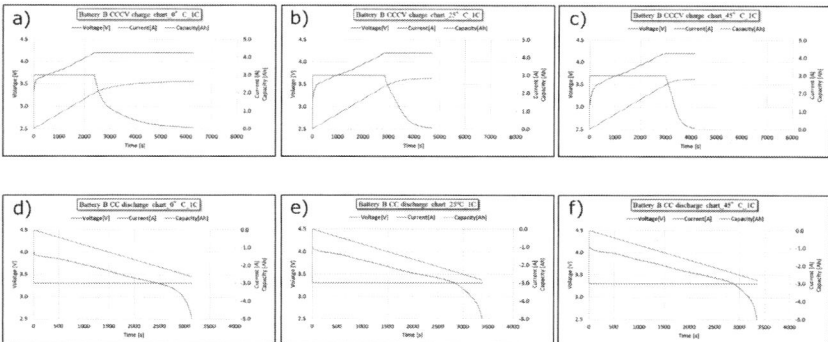

Fig.3 Charge and discharge characteristic curves of LiB-B at 0□, 25□ and 45□

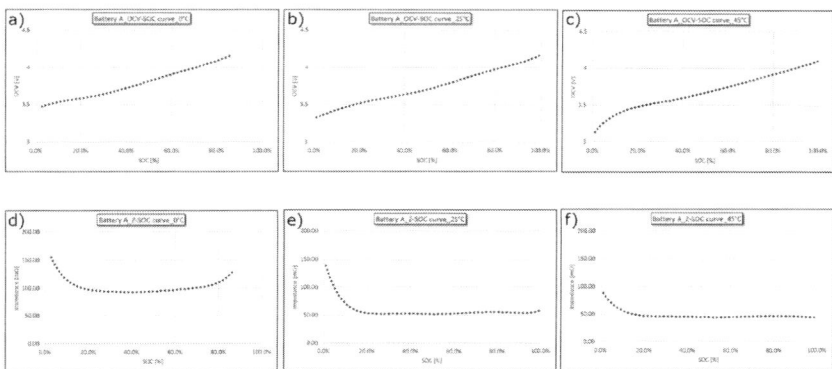

Fig.4 OCV / impedance and SOC lookup tables for LiB-A at 0□, 25□, and 45□

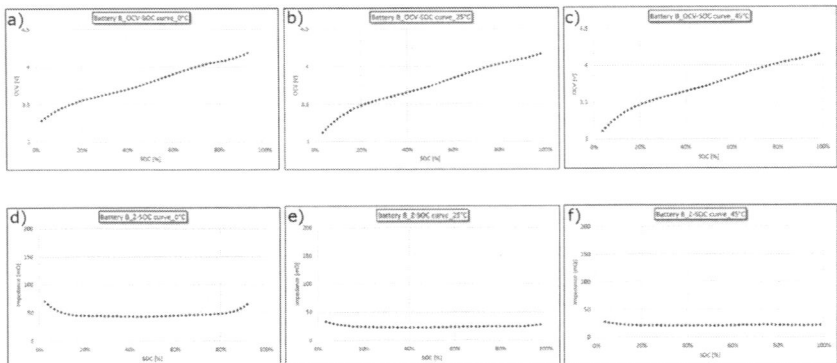

Fig.5 OCV / impedance and SOC lookup tables for LiB-B at 0□, 25□, and 45□

The chargeable capacity of a LiB greatly depends on the environmental temperature during charging. The chargeable capacity around the lowest and highest chargeable temperature may decrease by tens of percent as compared with that at nominal voltage charging temperature. Thus, charging at 25□, discharging at 0□, re-charging at 25□, discharging at 45□ were also measured in consideration with the initial FCC. These data ware be used in later simulation evaluation.

3 Modeling

Modeling was performed using the simplest systematics of a Foster-type equivalent circuit, referring to G.L. Plett's pioneering work [5] and as shown in Figure 6.

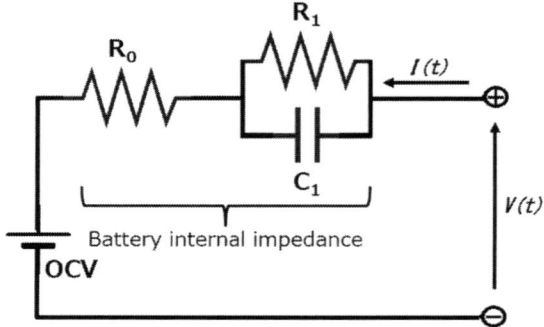

Fig.6 Foster-type equivalent circuit model

Here, R_0 denotes the resistance to mass transfer of lithium ions in the electrolyte; R_1 and C_1 denote resistance to charge transfer on the electrode surface, i.e., solvation and desolvation resistance, and electric double-layer capacitance respectively. The terminal voltage and current are defined as V(t) and I(t).

3.1 Current Integration Method

The current integration method uses formulae (1) and (2):

$$Q(t) = \int_0^t I(t)dt \cong \sum_{k=0}^n \triangle Q(k) \tag{1}$$

$$\triangle Q(k) = (I_k + I_{k+1}) \times \triangle t/2 \tag{2}$$

3.2 Extended Kalman filtering (EKF) method

Since the LiB system is nonlinear, a diagram of nonlinear discrete-time state-space considering noise is introduced by using EKF modeling, as shown in Figure 7. The mathematical expression is shown as formulae (3) and (4).

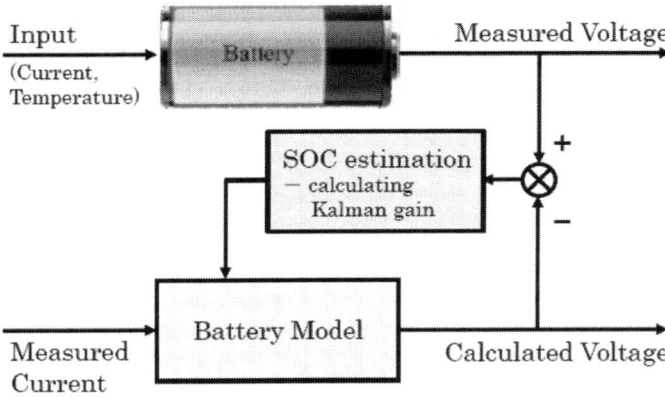

Fig.7 Diagram of nonlinear discrete-time state-space for EKF

$$y(k) = V_p(k) - V(k) \tag{3}$$

$$\widehat{SOC} = \widehat{SOC^-} + G \times y(k) \tag{4}$$

Here, $\widehat{SOC^-}$ and $\widehat{SOC}$ denote the prior SOC estimation and updated SOC estimation; V_p and V denote the measured voltage and the calculated voltage; y(k) denotes the error between measured voltage and calculated voltage; and G denotes the Kalman gain.

Figure 8 shows EKF implementation for SOC estimation on MATLAB® R2017b, with reference to the instructions from Adachi et al. [6].

```
function [xhat_new,P_new,G] = ekf(f,h,A,C,Q,R,y,xhat,P)
xhat=xhat(:);
y=y(:);

xhatm=f(xhat);
Pm=A(xhat)*P*A(xhat)' +Q;
G= Pm*C(xhatm)'/(C(xhatm)*Pm*C(xhatm)'+R);
xhat_new=xhatm+G*(y-h(xhatm));
P_new=(eye(size(A(xhat)))-G*C(xhatm))*Pm;

end
```

Fig.8 EKF implementation for SOC estimation on MATLAB

4   Simulation on LiB-A

Simulation has been performed on LiB-A for different algorithms using the current integration method and EKF modeling method for discharge at 0□, 25□ and 45□, respectively, inconsideration of simulation results shown in Figure 9.

The left side of Figure 9 shows simulation results for the current integration method. From these results, it is demonstrated that the fact that initial and measurement errors are perpetual can be considered to be one of the main characteristics of the current integration method.

On the other hand, the right side of Figure 9 shows simulation results for the EKF modeling method, which provides highly accurate SOC estimation values where the maximum absolute error is not only below 2%, but also converges in the decreasing direction.

Figure 10 proposes a process for quick SOC estimation.

5   Verification using LiB-B

Verification was performed on LiB-B by executing the process as mentioned in Figure 10, with different algorithms from the current integration method and the EKF modeling method, for discharge at 0□, 25□ and 45□, respectively.

Figure 11 shows the verification results which were quite similar to those of LiB-A. Therefore, the verification results themselves are not needed for detailed explanation, because the observations for LiB-A were reproduced also for LiB-B. An insightful view of this verification is that the proposed process for SOC estimation has been proved with high precision and high efficiency. Firstly, a very short time, normally 1 to 2 weeks, was taken to reach the breakpoint test; secondly, the computational complexity was significantly reduced due to the application of only known breakpoints to models; and finally, the estimation was reproducible and estimation error was comparable.

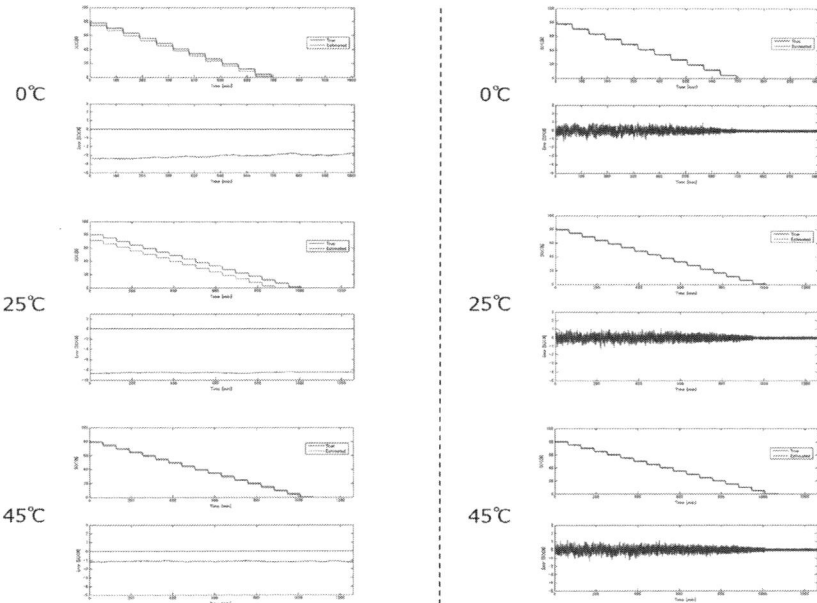

Fig.9 Simulation results on LiB-A for current integration method (left) and EKF modeling method (right) – at each temperature, upper: transition of SOC true value and SOC estimation value, lower: SOC error in [%]

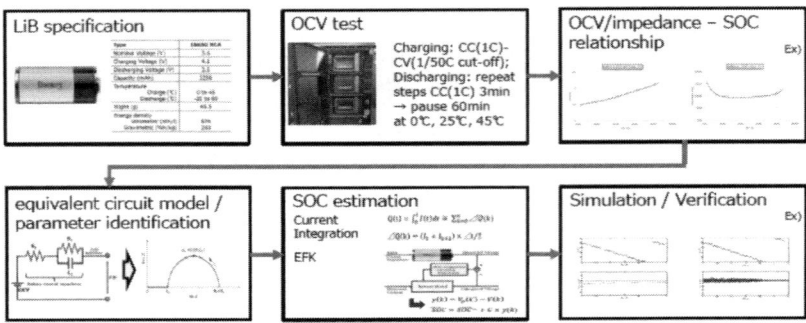

Fig. 10 A proposed process for SOC estimation

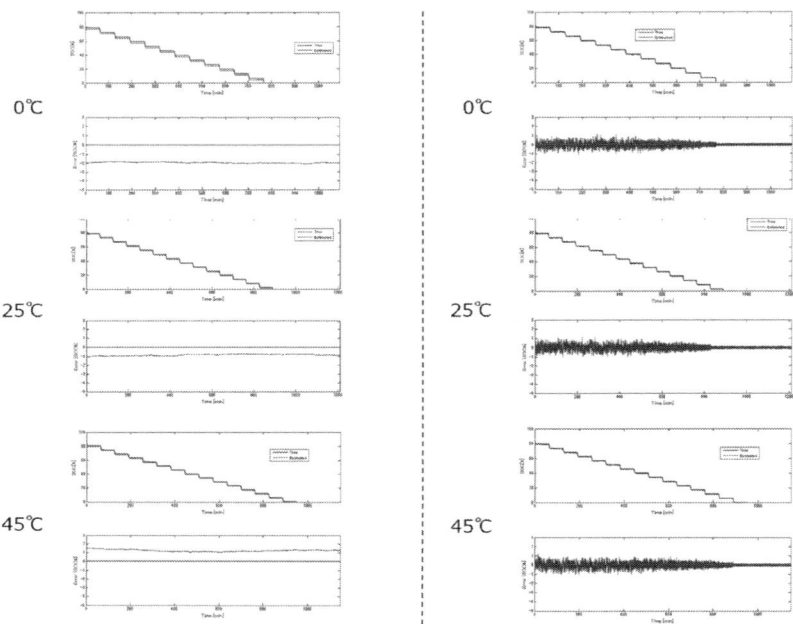

Fig.11 Verification results of LiB-B for current integration method (left) and EKF modeling method (right) – at each temperature, upper: transition of SOC true value and SOC estimation value, lower: SOC error in [%]

6 Summary

The following achievements were obtained in this study.

1. Appropriate and effective pretest procedures were used to determine the characteristics of the battery in a short time. Typical charging and discharging data from the measurements taken provided bases for creating OCV / impedance and SOC relation tables, and the models were subsequently identified without backtracking.
2. As mentioned in Chapter 1, each of the methods used for LiB capacity estimation have their inherent merits and demerits. The OCV method, the current integration method, and the EKF modeling method have been compared by the experiments conducted during this study. The modeling method using a Kalman Filter shows relatively high accuracy and error convergence within a certain time period.
3. A process for SOC estimation was proposed and performed on one type of LiB. The effectiveness of this process was verified on another type of LiB, showing its ability in common use.

However, factors such as battery deterioration and cell balancing that influence SOC estimation, have not been discussed in this paper, and simultaneous estimation of SOC and the parameters method based on the UKF (unscented Kalman filter) algorithm also have not been included in this study.

Future studies will cover the above and focus more on peculiar problems with batteries of EVs, especially adding comparative analyses using the NEDC (New European Driving Cycle), and the C-WTVC (Chinese version World Transient Vehicle Cycle) test.

References

1. Huang, F.L., Sumida, Y., Nomura, A., Matsumura, H., Kamiya, Y., Daisho, Y., Morita. K., "Analysis of adverse effects on vehicle performance due to hybrid vehicle battery deterioration," Proceedings of the 26th international electric vehicle symposium (EVS26)
2. Tsuchiya, S., "Integrated Battery Management System Combining Cell Voltage Sensor and Leakage Sensor," Keihin Technical Review, Vol. 6, pp. 62-67, 2017
3. Tsuchiya, S., Kamata, S., "Battery Voltage Detection Device," Japan Patent 2016-194428, November 17, 2016 (Request for Substantive Examination)
4. Charge/Discharge System Controller PFX2512 Catalog (Preliminary), http://www.kikusui.co.jp/catalog/pdf/files/2012/pfx2512.pdf, accessed Nov. 2017
5. Plett, G.L., "Extended Kalman filtering for battery management systems of LiPB-based HEV battery packs — part 1: Background, part 2: Modeling and identification, part 3: Parameter estimation," Journal of Power Sources 134 (2), 252-292, 2004
6. Adachi, S., Hirota, Y. (Eds.), Osiage, K., Baba, A., Maruta, I., Teruyoshi Mihara, T., "Battery Management Systems Engineering – From Battery Structure to State Estimation,"Tokyo Denki University Press, 2015

Authors

Fuliang Huang(Speaker)

Team Leader/Battery Management System
TEL: +81-28-680-1529
FAX: +81-28-680-1515
Email:fuliang-huang@keihin-corp.co.jp

F. Huang received a M.D. of engineering from Waseda University. He started at Keihin in the Hardware Development Department. He is now working as Team Leader of BMS control development in the Battery Management System Development Department.

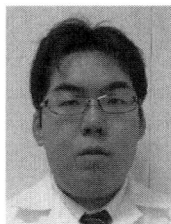

Masashi Murohoshi

Segment Leader/Battery Management System
TEL: +81-28-680-1529
FAX: +81-28-680-1515
Email:masashi-murohoshi@keihin-corp.co.jp

M. Murohoshi received a Bachelor of Science in Information Technology from Niigata University of International and Information Studies. He started at Keihin in the Software Development Department. He is now working as Segment Leader of BMS control development in the Battery Management System Development Department.

Akira Ichinose

Engineer/Battery Management System
TEL: +81-28-680-1529
FAX: +81-28-680-1515
Email:akira-ichinose@keihin-corp.co.jp

A. Ichinose received a M.D. of engineering from Tokyo Denki University. He started at Keihin in the Software Development Department. He is now working as an Engineer of BMS control development in the Battery Management System Development Department.

Tingting Sui

Engineer/Battery Management System
TEL: +81-28-680-1529
FAX: +81-28-680-1515
Email:zui-teitei @keihin-corp.co.jp

T. Sui received a M.D. of engineering from Fukuoka Institute of Technology University. She started at Keihin in the Hardware Development Department. She is now working as an Engineer of BMS control development in the Battery Management System Development Department.

A test center for aging analysis on Li-ion cells for automotive series application – test equipment, test procedures and cell aging effects

Peter Haußmann, Joachim Melbert

Introduction

Electric vehicles (EV) are a promising alternative to vehicles based on combustion engines: they allow for emission-free driving and eliminate the dependence on fossil fuels. However, the main limiting factor of contemporary EVs is energy storage, which usually consists of Lithium ion (Li-Ion) cells. In comparison to fuel tanks in combustion-based vehicles, today's Li-Ion batteries are heavy, require long recharging times and provide only a limited driving range. In addition, battery safety requires continuous battery monitoring in automotive applications. The most important difference, however, is the limited life time of the battery cells, which is affected by various aging effects. This problem is addressed in ongoing research by multiple approaches – from improved cell chemistries to optimized operating strategies. Every approach requires extensive pre-series testing to determine the best available cell type for the given application and to validate the cell life time under realistic boundary conditions. These tests are usually performed on single battery cells rather than battery packs in order to minimize the testing effort.

This work is structured into four parts: In the first, different experiment strategies for cell aging analysis are discussed, which are designed to analyze either distinct aging effects or the complex interactions of multiple cycling parameters. Both types of experiments consist of repetitive cycling and characterization periods.

The second part describes a single-cell precision test equipment. It is optimized for both high-current cycling and accurate cell characterization. The test center is equipped with more than 150 cell testers and climate chambers. It is organized in a modular concept. A dedicated safety and infrastructure supervision concept enables scheduled long-term experiments with high uptime.

In the third part, aging effects measured in different life time experiments provide an insight into the complex interactions of various aging mechanisms in Li-Ion cells. Standard cell properties such as capacity and internal resistance are observed to determine the cell aging state. In addition, further aging effects of the open circuit voltage and cell impedance are shown.

The last part of this work provides an insight into advanced measurement methods for Li-Ion cells and applications based on improved cell modeling. Concepts for self-discharge detection and internal cell temperature measurement without dedicated sensors are presented. Detailed modeling data sets also enable cell behavior simulation and hardware-in-the-loop cell emulation. These techniques supplement the development and testing of additional EV components, such as battery management systems.

Aging Effects on Automotive Li-Ion Cells

The electrical battery represents the most critical part of electrical vehicles concerning cruising range, acceleration, safety aspects and, most notably, aging. Li-Ion technology and its various derivatives represent the state of the art. Actual development is focused on increased specific capacity, with remaining safety and robustness. Before entering series application, new cells must be fully characterized. Serious contradiction results from the intended lifetime of 8 years and the much shorter time for the proof of usability.

EV batteries lose capacity due to various aging effects, which reduces the maximum driving range for a full battery charge. The remaining driving range until the battery must be recharged is given by the cell state of charge (SoC). It is defined as the ratio of the remaining amount of charge in the cell and the actual cell capacity.

Cell aging is usually described as capacity degradation. A standard convention for the cell state of health (SoH) allows 20 % fade of capacity before the cell reaches its end of life (EoL) state. Also the peak power capability is affected by cell aging. Increase of the internal cell resistance reduces the maximum available momentary power. This must also be considered in EV energy storage concepts.

A large variety of cell formats and chemistries is available for EV applications. The specified nominal capacity must be chosen appropriately for the target application. However, aging effects can differ significantly for cells with equally rated capacity. Figure 1 shows an example of capacity and internal resistance aging for two cells with similar chemistries which were cycled using identical profiles. The observed aging effects differ completely, although the differences in material composition were declared as minor change.

Aging Effects for different Cell Chemistries

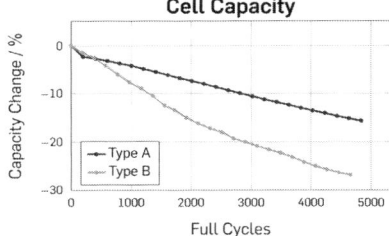 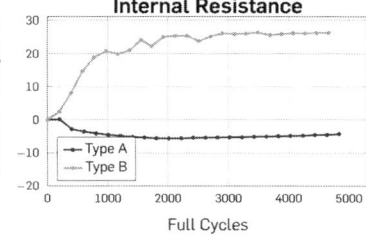

Figure 1: Capacity and Internal Resistance Aging Effects for different Cell Chemistries.

3

For a given cell type, different types of aging effects must be considered, which can be classified in two categories: **Calendar aging** refers to aging effects caused by side-reactions during rest and storage periods. It is described as a function of time. **Cyclic aging** is introduced during cycling operation. It is described as a function of charge throughput. Both categories are influenced by multiple operating parameters:

Temperature

State of charge SoC

Depth of discharge DoD

Current profile

In Figure 2 aging effects for two different cycling profiles of one cell type are compared. All operation parameters except for the SoC operating window are identical for both profiles. Yet, substantially different aging effects are observed. Without precise knowledge of the individual parameter influences and their interaction on cell aging, cell life time cannot be sufficiently determined.

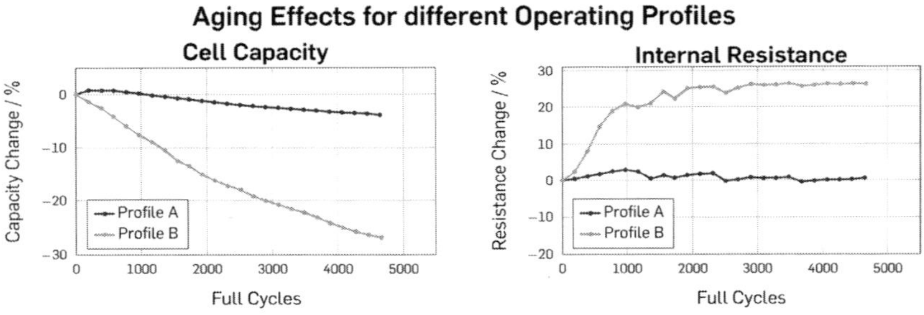

Figure 2: Capacity and Internal Resistance Aging Effects for different Cycling Profiles.

Aging Analysis

Most of the stimulation parameters show nonlinear aging results. With respect to the various individual aging effects and their complex interaction, different types of experiments can be applied.

Analysis of single-parameter influences provides a better understanding of the aging impact for individual parameters. Often synthetic cycling profiles are applied as they provide dedicated excitation characteristics. However, the results are only valid for a small operating area, described by the other fixed parameters.

Statistical aging studies consider influences of several parameters as well as their interactions. The results are used for parametrization of cell aging models. The analysis of multiple parameter constellations is restricted, as it requires a large number of test cells. Statistical design of experiment (DoE) methods reduce the number of cells while minimizing information loss [1].

Real driving and temperature profiles are used to validate the determined aging models under realistic operating conditions. They are also suited for benchmarks of several cell types to identify the optimal cell technology for a given application.

The application of an automotive battery can have usage time of up to one decade. Such a long period is not available for pre-series testing. Therefore, aging experiments must be designed to observe all relevant aging effects at a minimum test duration without altering the cell operation profile. Due to the nonlinear parameter influence, current or temperature increase for acceleration is not feasible. Instead, long rest periods between cycles can be avoided and calendar aging must be considered. The charge throughput of an ordinary 5-10-year customer application under realistic driving conditions can thereby be reproduced in approximately 1 year of cycling.

Such test procedures consist of **cycling phases** to stimulate aging and of **characterization phases** to evaluate aging effects of typically 100 to 150 cells. The aging effects must exclusively be stimulated during the cycling phases. The characterization duration must be kept as short as possible to avoid undesired recovery effects and the charge throughput must be kept as small as possible. Figure 3 shows a schedule applied in many studies. Alternating 14-day cycling phases and 1-day short characterizations (SC) achieve a good balance. High-detail characterizations including open circuit voltage (OCV) and electrical impedance spectroscopy (EIS) measurements at multiple temperatures provide in-depth modeling data for the investigated cell types. These procedures are conducted at the beginning and at the end of the study. Additional modeling data for more aging states can optionally be extracted in extended characterizations every 3 to 6 months.

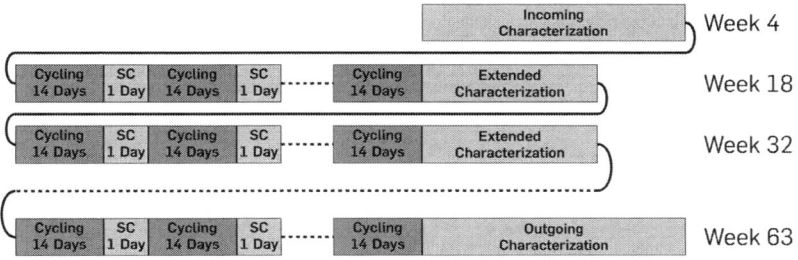

Figure 3: Time Schedule for Aging Studies.

5

Although cell aging distribution for identical test procedures is quite low, all tests are performed on multiple pre-selected cells with minimum capacity and mass deviation. Using the described method, high performance for several cell types of various technologies, capacities and test procedures could be proven.

Cell Test Equipment

The realization of cell aging studies as proposed requires appropriate test equipment to provide both high-current cycling for aging simulation and high-accuracy characterization at highest reliability. For this purpose, a compact 19" single-cell test unit was developed [2]. It uses a hybrid power stage concept: a linear wide-bandwidth high-current current sink is capable of high discharge current at low rise times. Charge current is provided by an external high efficiency switched-mode supply. The combination can reproduce dynamic high-current automotive cycling and recuperation profiles with low output ripple. The test system includes a precision measurement unit to acquire cell current, voltage, and temperature at high sampling rate and resolution. For higher power, the units can be paralleled. Table 1 gives an overview of the system specifications.

Table 1: Cell Test System Specifications.

Power Stage Specification	
Discharge Current Range	0 – 600 A
Charge Current Range	0 – 200 A
Rise Time	80 μs
Sample Output Rate	100 kHz
Voltage range	0.2 – 5 V
Max. Discharge Power	1.8 kW
Measurement Unit Specification	
Voltage Accuracy	50 μV
Current Accuracy	0.1 % (full range) [3]
Temperature Accuracy	0.1 °C
Sampling Rate	50 kHz per channel
Resolution	18 Bit

An integrated control unit coordinates the power stage and data acquisition unit. Various operating modes are supported such as constant current (CC), constant power (CP) and constant voltage (CV) charge/discharge procedures. In addition, user-programmable current and power profiles as well as SoC tracking and control are available. The wide-bandwidth concept also allows for the generation of sinusoidal and arbitrary pulse waveforms with high signal integrity in a wide frequency range. For example, such signals are applied for pulse impedance spectroscopy, as described in [4, 5].

Cell voltage, current and temperature as well as system diagnostics are continuously monitored. Failure detection in hardware and software ensures safe and reliable operation.

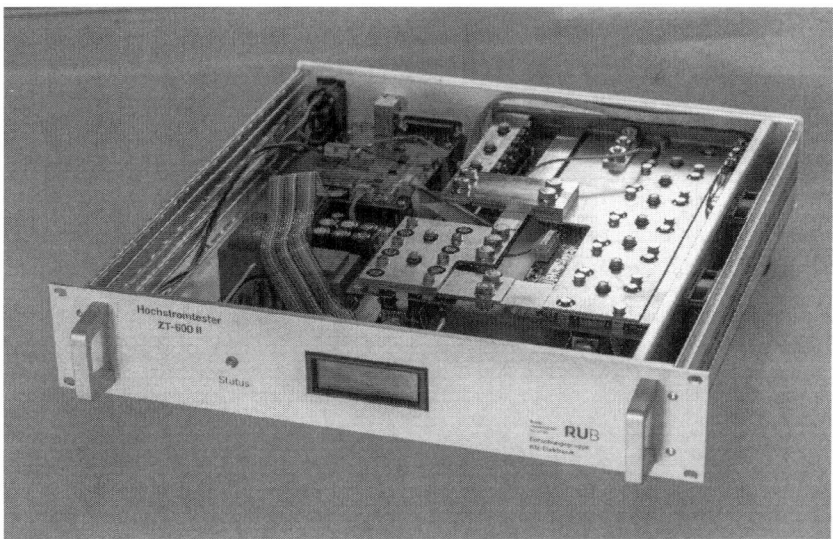

Figure 4: 600 A Cell-Test System

Cell Test Center

A test center for a maximum of 150 cells was realized and optimized for uninterrupted long-term aging studies under realistic automotive operating conditions. It is based on a modular structure and consists of 10 test compounds, as shown in Figure 5.

Each test compound consists of up to 18 cell test units and an associated temperature chamber to control the ambient temperature during the tests. Water cooling with individual temperature regulation for each cell allows for simulation of the cooling concept of automotive battery modules and packs. One specialized supervision unit per compound monitors ambient temperature, power grid and water cooling integrity and other safety-relevant parameters.

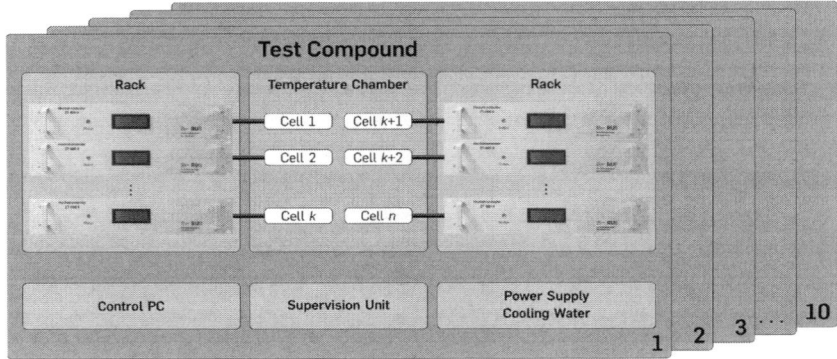

Figure 5: Structure of the Cell Test Center.

The battery tests are coordinated by a PC system running specialized software for test control, programming, scheduling, and measurement data storage. Until today, the cumulative charge throughput for cell aging experiments exceeds 30.000.000 Ah. The supervision concept achieves test center uptime of more than 99.9 %.

Figure 6: Left: Compound of 18 PHEV Cells in a Temperature Chamber. Right: Partial View on the Cell Test Center.

Cell Aging Effects

Characterization Methods

Aging effects are observed in characterization measurements. For reproducibility, the corresponding methods require well-defined parameters such as temperature, current rate, and the duration of rest periods.

Cell capacity is measured by charge counting during a standardized cell discharge process at constant current and temperature.

Internal resistance is obtained in current pulse experiments as the ratio of voltage and current variation after a defined current duration.

Temperature during cycling is measured to detect safety- critical changes of the cell behavior and for thermal cell modeling.

The information provided by capacity and internal resistance is sufficient for SoH analysis, but allows only limited modeling of the cell behavior at different temperatures and SoHs. Advanced characterization procedures provide an extended insight into the electrochemical cell properties:

Open circuit voltage (OCV) measurements analyze the relation between SoC and the idle cell voltage in the entire cell operating window. The results are relevant for several applications, such as SoC estimation algorithms [6].

Electrochemical impedance spectroscopy (EIS) is applied to measure the complex cell impedance for a wide range of frequencies. Resulting impedance spectra contain information about multiple electrochemical effects within the cell. The characteristic time constants of the observed effects differ significantly. Thus, they affect the cell impedance spectra in separate frequency regions. This enables to relate observed cell aging effects to distinct electrochemical cell properties.

OCV and EIS measurements usually require long measurement durations. Therefore, advanced procedures are developed to reduce measurement time and charge throughput.

Acceleration of Cyclic Aging Effects

Aging effects in test procedures must be stimulated faster than they occur in real applications. As mentioned above, this is achieved by shorter rest periods between the cycles. A proof of this approach was established in a separate study. A set of 8 PHEV cells with a rated capacity of 20 Ah was chosen and tested for 48 weeks using 4 different test programs:

1. Reference test without acceleration: a real 1-hour driving profile and a subsequent recharging process is applied once per day.
2. Acceleration 3x: the reference test profile is applied 3 times per day.
3. Acceleration 5x: the reference test profile is applied 5 times per day.
4. Calendar aging: no cycling; cells are stored at the average cycling SoC.

The total capacity aging measured for the first 3 test groups is shown in a diagram of Figure 7 (left) as a function of charge throughput. The dotted lines show linear extrapolation of the aging traces. For each group, the measured capacity reduction per charge throughput is different.

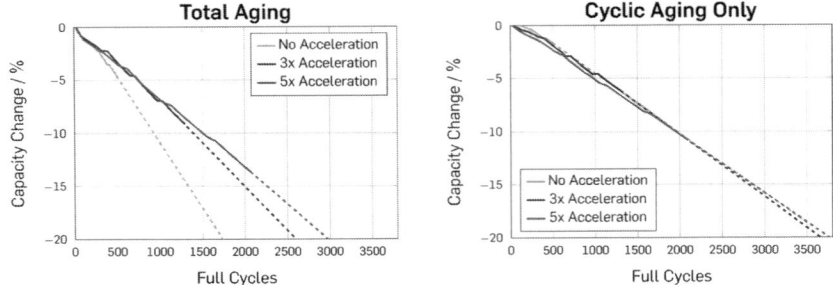

Figure 7: Validation Experiment for Aging Acceleration.
Left: Total Aging; right: Cyclic Aging only.

The different aging rates are caused by calendar aging effects. This is demonstrated in another diagram of Figure 7 (right), where the calendar capacity reduction is compensated for all tests. Thus, only cyclic aging effects remain. The extrapolations for all three tests are in good accordance. The results confirm that the proposed acceleration method does not alter the profile-specific cyclic aging rate.

State of Charge Operating Window

In the next procedure, the relevance of the mean state of charge (SoC) and depth of discharge (DoD) is analyzed in detail. In total, 14 different tests are structured into two test groups for 10 % and 20 % DoD, respectively. Within each group, the mean SoC is varied stepwise between 15 % and 85 %. Synthetic constant current profiles are used to provide distinct aging stimulation. In a 48-week experiment, 28 cells of a 20 Ah PHEV cell type were observed at T=25°C.

The measured capacity reduction for the 14 tests is shown by the horizontal bars in Figure 8. The horizontal bar range represents the SoC operating window, which is defined by the mean SoC and DoD. The vertical position indicates the total resulting capacity change. A substantial influence of the SoC cycling window is observed. Above 50 % SoC, severe aging effects exceeding the EoL criterion of 20 % capacity reduction are observed. High DoD leads to even more pronounced aging effects. The aging linearly decreases for SoC lower than 50 %. In this region, only the average SoC contributes to aging, regardless of the DoD.

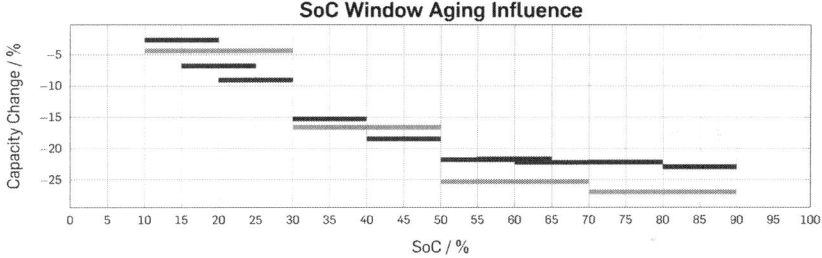

Figure 8: Capacity Aging for different SoC Operating Windows after 48 Weeks.

These results can be applied to optimize the operating strategy for EV and (P)HEV applications. For maximum driving range, high SoCs cannot be completely avoided. Yet, a longer life time is achievable if the cell is frequently operated in the SoC range below 50 %. For this, the battery should not be recharged too early.

Statistical Aging Study

The previously discussed experiment only considers variations of a few distinct parameters. Therefore, the obtained results are only valid under the boundary conditions defined by the constant experiment parameters, such as temperature and current profile. This can be overcome in statistical aging studies that analyze the aging contribution of multiple parameters and their interactions.

Aging effects for 120 PHEV cells with a rated capacity of 26 Ah are observed in a 48-week aging study. Unique real driving profiles were chosen for each cell based on a DoE approach. The profiles cover six variable parameters, which are listed in Table 2 with their corresponding range of values.

Table 2: Aging Study Parameters and Value Range.

Parameter	Minimum Value	Maximum Value
Discharge current (RMS)	0.3 C	3.5 C
Charge current (RMS)	0.3 C	1.5 C
Temperature	10 °C	40 °C
Mean Cycling SoC	26 %	91 %
Depth of Discharge	1 %	70 %
Charge Throughput	750 full cycles	3000 full cycles

11

Figure 9 shows the measured change in capacity and internal resistance during 48 cycling weeks. Grey areas represent the total range of the measured values for all 120 cells. Typical aging traces are included to show aging trends for cells with low, moderate, and increased aging effects.

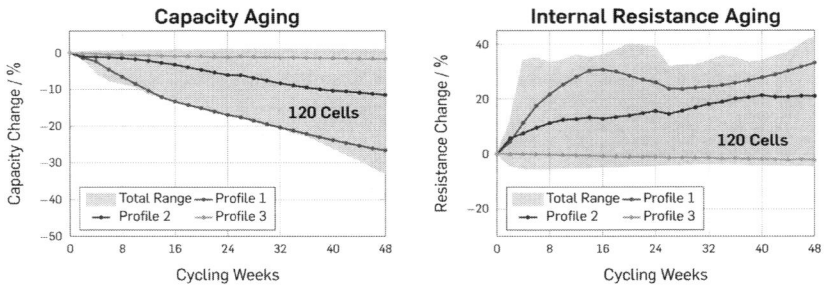

Figure 9: Measured Aging Effects for 120 Cells in a Statistical Aging Study. Left: Capacity; right: Internal Resistance. The grey Area presents the total observed Range for all 120 cells.

For the 120 cells tested the observed aging effects differ substantially. Aging effects caused by the parameters of Table 2 cannot be determined from single cells. The aging contribution of each parameter depends on the influence of the remaining parameters. This can be described by a cell aging model considering parameter interactions. The model coefficients are evaluated from aging data for all 120 cells to describe the complex interdependence. Figure 10 shows the local aging variation for all parameters in two different operating points.

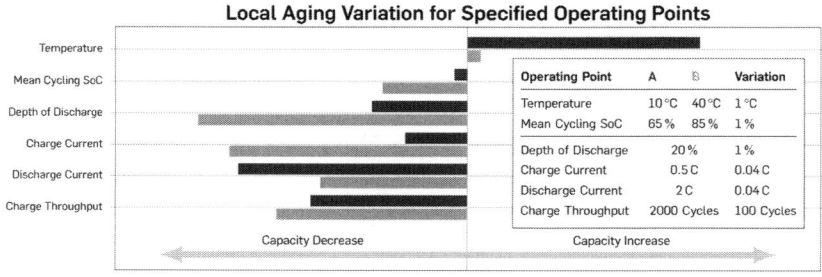

Figure 10: Local Aging Contributions of the Study Parameters for different Operating Points.

These operating points differ only in temperature and mean cycling SoC. All other parameters are identical. The local aging contribution caused by each parameter is determined separately. For this, the parameters are varied as specified in Figure 10 and the aging impact on the capacity is calculated. The determined influence on capacity fade differs significantly for the two operating points.

Changes in capacity and internal resistance directly influence the driving conditions. Other cell aging effects have an impact on the battery management unit and the control functions. Figure 11 shows aging effects for the open circuit voltage OCV and the cell impedance. Both effects are extracted from the 120 cell modeling test. During 48 cycling weeks, notable OCV and impedance changes are observed. These observations are used for high-detail cell behavior modeling over the complete lifetime. The generated data sets are suitable for advanced methods and algorithms; examples are discussed in the following section.

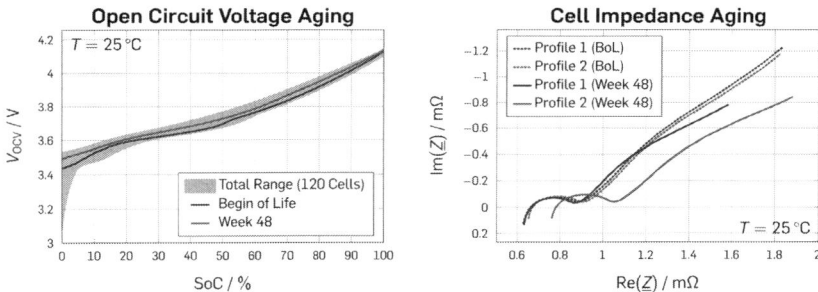

Figure 11: Aging of Cell Modeling Parameters. Left: Open Circuit Voltage; right: Cell Impedance.

Advanced Methods and Applications

Optimized Impedance Spectroscopy

EIS measurements provide detailed information about local electrochemical aging effects in Li-Ion cells. The cell impedance is sensitive to multiple parameters, such as temperature, SoC, and SoH.

Standard EIS methodology is mostly based on sinusoidal excitation, which contains only a single discrete frequency component. Impedance measurements for a wide range of frequencies and high resolution lead to a long measurement duration and high charge throughput. The test is widely used for laboratory purposes on small test probes.

Optimized EIS methods can be used to significantly reduce measurement duration and charge throughput. Instead of sinusoidal excitation, the cell is stimulated by wide-bandwidth pulse signals. Hence, the impedance is simultaneously analyzed across a wide frequency continuum rather than a single frequency component. A method optimized for short measurement durations and a high signal-to-noise ratio is presented in [4, 5]. The achievable improvements are shown in Figure 12. Despite a substantial reduction of time and charge throughput the penalty on data quality is low. This advanced method can now be included in the short characterization of Figure 3. As shown by further research EIS can be stimulated from real driving excitation and can be incorporated in future battery management systems.

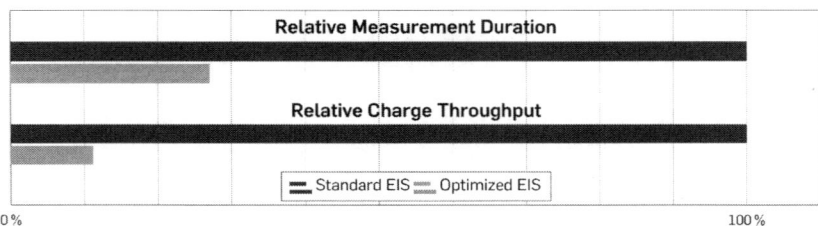

Figure 12: Reduction in Measurement Duration and Charge Throughput for the optimized EIS Method.

Self-Discharge Monitoring

Severe aging effects in Li-Ion cells can cause dendrite growth and increase the risk of an internal cell short circuit, which would impose a substantial safety risk. In the future, increased self-discharge will be monitored as an early warning indicator. For this purpose, the cell voltage is observed during relaxation periods. After the cell voltage has completely settled from the previous cycling load, self-discharge becomes manifest as a continuous voltage decrease. This behavior is depicted in diagrams of Figure 13 (left) for both a healthy and a high-leakage cell.

The measured long-term voltage gradient is proportional to the leakage current in the cell. The main disadvantage of this method is the required duration of the rest period of several hours or even days. Faster self-discharge detection can be achieved by means of model-based or statistical approaches [7]. Diagrams of Figure 13 (right) show statistical self-discharge indicator values for 120 cells. The distribution was evaluated during a statistical aging study after 24 and 48 weeks, respectively. The high resolution and low drift of the described cell tester allows for leakage analysis without any dedicated equipment during the long-term aging tests.

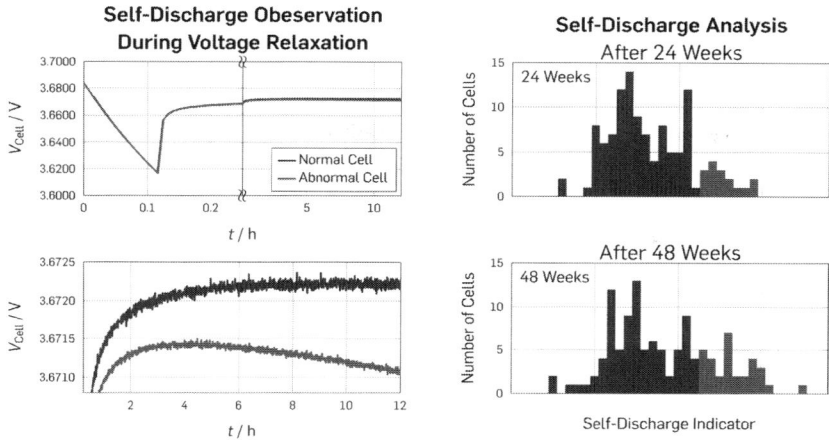

Figure 13: Self-Discharge Measurement by Voltage Gradient Analysis. Left: measured Voltage Gradient; right: Statistics-based Self-Discharge Indicator Distribution for 120 Cells at different Aging states.

Internal Cell Temperature Measurement

The internal temperature of Li-Ion cells is a safety-relevant operation parameter. However, temperature measurement using standard sensors is only possible at the cell surface. For lower cost and complexity, temperature is often restricted to a single sensor per cell module. The distinct temperature dependence of the electrochemical cell properties can be used to measure the internal cell temperature without dedicated sensors. The diagram of Figure 14 (left) shows cell impedance for different temperatures. This relation is applied in a sensorless measurement technique for the internal cell temperature based on the common cell voltage and current measurement in a standard battery management unit [8].

The method is demonstrated in the right part of Figure 14 where a real driving cycle is used for excitation. For comparison, the cell surface temperature measured by a standard sensor is shown. The internal cell temperature exceeds the cell surface temperature by several °C. Using the proposed approach any critical thermal cell state can be detected early, and substantial improvements in battery safety monitoring are achievable by the method.

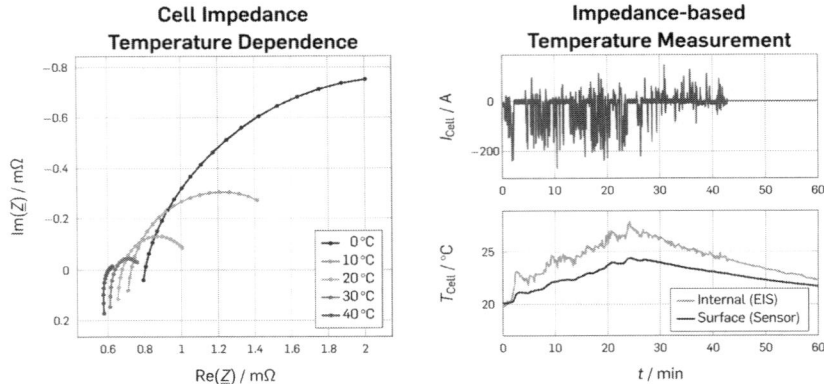

Figure 14: Internal Cell Temperature Measurement by Means of Impedance Spectroscopy. Left: Impedance Temperature Dependence; right: Temperature Response for Driving Profile.

Hardware in the Loop Testing of Battery Management Systems

For optimal battery performance and life time, battery management systems (BMS) monitor the condition of each cell in automotive energy storages. A BMS also applies cell balancing to compensate for different cell SoCs within the battery pack. The ongoing development of BMS hardware and software allows for advanced diagnostic capabilities and algorithms in future BMS. Prior to series application these new features must be thoroughly evaluated in order to validate functionality for all relevant battery operating modes and SoHs. BMS tests using real battery packs are inflexible, time consuming and imply higher risks. The performance of the system results from a proven cell model which must be based on extensive tests.

Hardware-in-the-loop (HiL) testing of BMS systems uses cell emulation devices instead of real battery cells. A high-performance single-cell emulation unit was developed, which is capable of accurate wide-bandwidth behavior simulation [9]. The concept is supported by the broad and accurate modeling data sets measured during aging studies.

The emulation capabilities are demonstrated in Figure 15. Results of impedance measurements for both the emulated cell type and the emulator are in good accordance, as can be seen in the diagram (left). The diagrams on the right-hand side show measurement data of a 1-hour charging process in an active balancing procedure, which is measured for a real cell and the emulation system. At time scales ranging from milliseconds to hours, both measurements are indistinguishable.

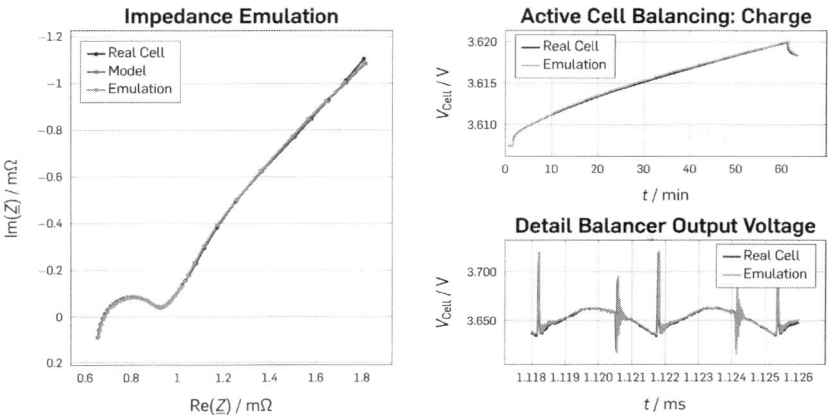

Figure 15: Demonstration of Cell Behavior Emulation in HiL Application. Left: Impedance Spectroscopy Measurement; right: Waveform for Active Cell Balancing.

Such a novel cell emulator incorporates attributes of cell test, cell modeling, circuit design and fast real-time software. The system improves the validity of HiL tests and enables the development of novel advanced BMS functionality, e.g. the previously discussed safety monitoring approaches. The advanced HiL testing concept offers a more rapid development of BMS systems and algorithms.

Conclusions

Continuous improvement of Li-Ion cells concerning energy density, aging and safety aspects require extensive analyses prior to series production of automobiles. Confirming operation times for the customer which are much longer than the time required for characterization of the cells is challenging. In this work, different types of characterization and aging experiments for analysis on Li-Ion cells are discussed. Specialized test equipment is presented which meets the demands for both realistic high-current cycling and accurate characterization measurements. The shown test center for 150 EV cells is suitable for large-scale aging analysis on Li-Ion cells with high test flexibility and up-time. The current range covers the whole spectrum of present and foreseeable cell sizes. All test objects are equipped with an individual temperature control in order to reproduce their ambient conditions of application in cars.

Results of different long-term aging experiments were presented. The aging effects are documented by the cell capacity and internal resistance change. Aging data show both the impact of distinct cycling parameters and the complex interaction of multiple parameters. Further cell properties can be observed by precision measurements of the open circuit voltage and cell impedance in multiple aging states. Advanced modeling including such effects enables new features such as sensorless temperature monitoring, leakage detection or SoC and SoH prediction including aging effects. Such techniques can improve safety and reliability of the battery with minimum effort in existing battery management architectures.

Extended cell modeling together with advanced circuit design leads to a new cell emulator for hardware in the loop application. The static and dynamic performance was demonstrated by several tests such as impedance spectroscopy and even active balancing.

The presented test center, the test equipment and algorithm concept meet all requirements on pre-series cell testing. It is used for aging analysis and the definition of advanced Li-Ion cell algorithms and applications in close cooperation with the electric vehicle manufacturing industry.

References

1. Fedorov, V.V., "Theory of optimal experiments," Probability and mathematical statistics, Academic Press, New York, ISBN 9780323162463, 1972.
2. Weßkamp, P., Haußmann, P., and Melbert, J., "600-A Test System for Aging Analysis of Automotive Lithium Ion Cells With High Resolution and Wide Bandwidth," *IEEE Trans. Instrum. Meas.* 65(7):1651–1660, 2016, doi:10.1109/TIM.2016.2534379.
3. Weßkamp, P. and Melbert, J., "High-accuracy current measurement with low-cost shunts by means of dynamic error correction," *J. Sens. Sens. Syst.* 5(2):389–400, 2016, doi:10.5194/jsss-5-389-2016.
4. Haußmann, P. and Melbert, J., "Spannungsgeregelte Impedanzspektroskopie mit breitbandigen Anregungssignalen für Lithium-Ionen-Zellen in Kfz-Anwendungen," *tm – Technisches Messen* 84(6), 2017, doi:10.1515/teme-2017-0018.
5. Haußmann, P. and Melbert, J., "Optimized mixed-domain signal synthesis for broadband impedance spectroscopy measurements on lithium ion cells for automotive applications," *J. Sens. Sens. Syst.* 6(1):65–76, 2017, doi:10.5194/jsss-6-65-2017.
6. Wesskamp, P., Reitemeyer, S., and Melbert, J., "Online Capacity Estimation for Automotive Lithium-Ion Cells Incorporating Temperature-Variation and Cell-Aging," SAE Technical Paper Series, WCX™ 17: SAE World Congress Experience, APR. 04, 2017, SAE International400 Commonwealth Drive, Warrendale, PA, United States, 2017.
7. Haussmann, P. and Melbert, J., "Self-Discharge Observation for On Board Safety Monitoring of Automotive Lithium Ion Cells – Aging Effects and Accelerated Procedures," in: *SAE World Congress Experience 2018, under review.*
8. Haussmann, P. and Melbert, J., "Internal Cell Temperature Measurement and Thermal Modeling of Lithium Ion Cells for Automotive Applications by Means of Electrochemical Impedance Spectroscopy," *SAE Int. J. Alt. Power.* 6(2), 2017.
9. Lueke, C., Haussmann, P., and Melbert, J., "A Modular Wide Bandwidth High Performance Automotive Lithium-Ion Cell Emulator for Hardware in the Loop Application," in: *SAE World Congress Experience 2018, under review.*

Development methods for RDE-compliant powertrains

Dipl.-Ing. Tobias Mink, Dipl.-Ing. Christian Lensch-Franzen,
Dipl.-Ing. Martin Schäfer, Alexander Ebel, M.Sc.

© Springer Fachmedien Wiesbaden GmbH, ein Teil von Springer Nature 2018
J. Liebl (Hrsg.), *Der Antrieb von morgen 2018*, Proceedings,
https://doi.org/10.1007/978-3-658-21419-7_8

Abstract

The ongoing change in mobility philosophy and the increased awareness regarding emission influence create new challenges for the powertrain development with focus on efficiency enhancement and emission reduction, especially under real driving conditions. As a result, the emission behaviour shows a high significance in addition to drivability, taking into account the overall lifetime and the increasing spread of load and gradient. At the same time, development cycles are becoming shorter and development costs are rising due to higher complexity and number of derivatives. This requires a systematic and efficient approach early in the powertrain development phase. To achieve the ecological and economic development objectives within shorter product cycles, the diversity of different influencing factors must be considered, analysed and, if necessary weighted according to their importance. This requires the use of a suitable development methodology, including target-oriented simulation tools, dynamic measurement techniques and a profound understanding of physical phenomena and mechanisms.

In order to combine the increasingly complex interactions and the upcoming development targets, APL has developed a method chain based on a real driving data base (APL TrackKit) with different powertrain setups. The collected data is classified in terms of combustion process, specific power, transmission spread and vehicle-specific driving resistances. Based on a Design of Experiments (DoE) approach, a target configuration for the future powertrain setup is derived, considering the main influencing factors such as vehicle, driver, route and environmental parameters are identified. A systematic parameter variation of hardware, operating strategy and calibration is performed on the test bench, using a hardware-in-the-loop configuration with a real time simulation environment. The main optimization targets are energy efficiency, limited emissions and driving behaviour.

The combination of state-of-the-art engine test benches, a simulated vehicle environment and online-tools as well as approval-relevant exhaust gas measurement technology of the RDE legislation enables optimal calibration and specific selection of hardware components in relation to the RDE-development both in the early development phase and during development.

1 Future challenges of powertrain development

The target is the development of powertrain configurations, which allow a technically maximum possible reduction of tailpipe emission under real driving conditions. The resulting development tasks for the automotive industry are comprehensible and cannot be mastered without simulation-based methods, which accompany the complete development process parallel to test cycles. APL has developed a methodological chain that integrates into the entire development process, taking into account the above-mentioned

2

targets, which enables an early assessment with regard to emission sensitivities and thus supports the path to an emission-optimized powertrain (Figure 1). The RDE optimization process determines the main influencing factors of the target configuration using real-time simulation, hardware-in-the-loop configurations and vehicle validation.

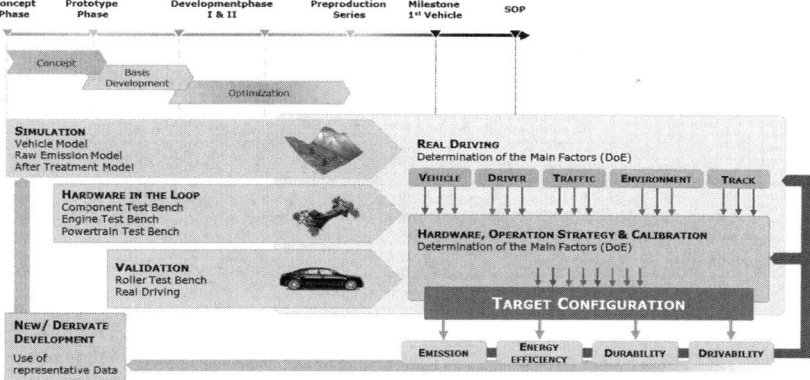

Figure 1: Optimized vehicle development process under RDE boundary conditions

In the following, this method chain is presented by means of an emission assessment over vehicle lifetime taking into account wear behaviour as well as its optimization. Emission stability over running time is of particular interest in connection with aging mechanisms of exhaust gas aftertreatment components such as three-way catalysts and particle filters. For this reason, engine-out emission reduction plays a central role, which can only be achieved by means of a consistent optimization in the interaction of engine mechanics and combustion process. An estimation of the emission degradation over wear phenomena can thus be used in advance for a better evaluation of the engineering target.

2 General approach for development

The general development approach for efficiency and emission optimization under real driving conditions presented in Figure 2 is based on the selective modelling of various driving conditions. The aim of the methodology is to build up an essential physical system understanding already in the early phase of a powertrain development, in order to obtain estimates about the emission levels under real driving conditions. Driving cycles and the load maps resulting therefrom can either be synthesized from the APL database or derived from real driving data generated by existing powertrains from previous generations.

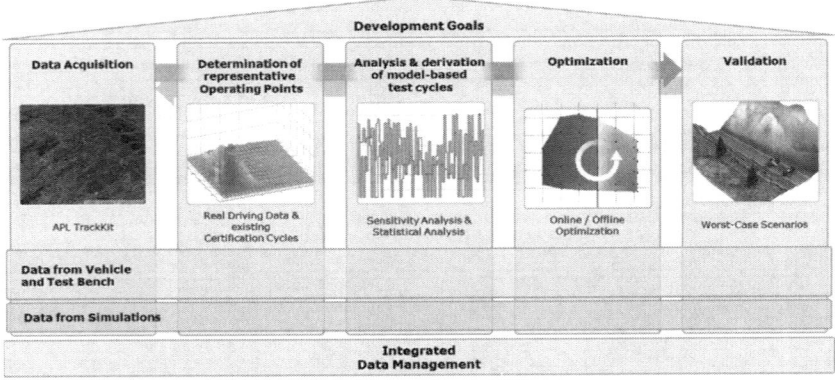

Figure 2: General development approach – overview of methodology

As a basis for the data acquisition of real driving data, a selection of routes around Landau with various load profiles is used. The APL TrackKit (Figure 3), which is used therefore, contains 32 individual modules that are connected to each other at 12 junctions.

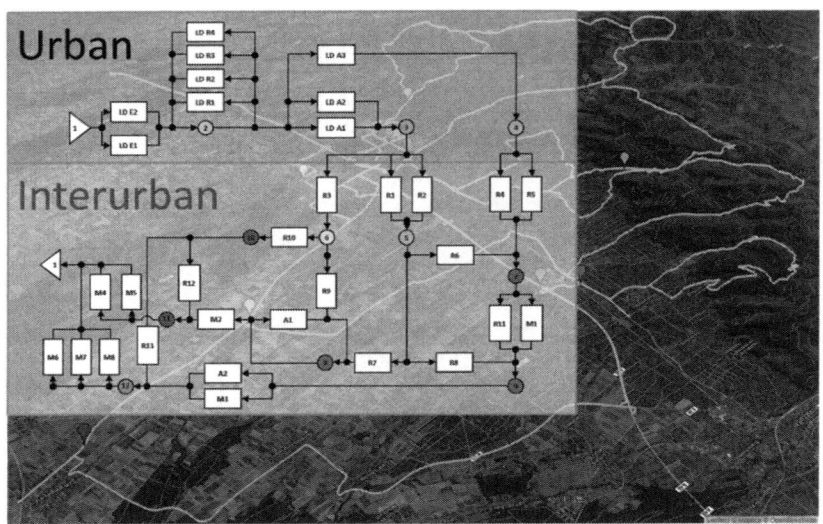

Figure 3: APL TrackKit – selection of route sections

The RDE online toolset allows online monitoring of the complex boundary conditions as well as the influence of the dynamic and the driver to validate the RDE measurements during the measurement run. Furthermore, measured emissions are taken into account, for example to use routes or driver behaviour with tendency to high emissions for worst-case validation. As a result, a plurality of different RDE-compliant rounds with a low error rate can be performed.

A simulation environment (Figure 4) enables the optimization on the engine test bench, powertrain test bench and roller test bench under repeatable and controlled conditions [4].

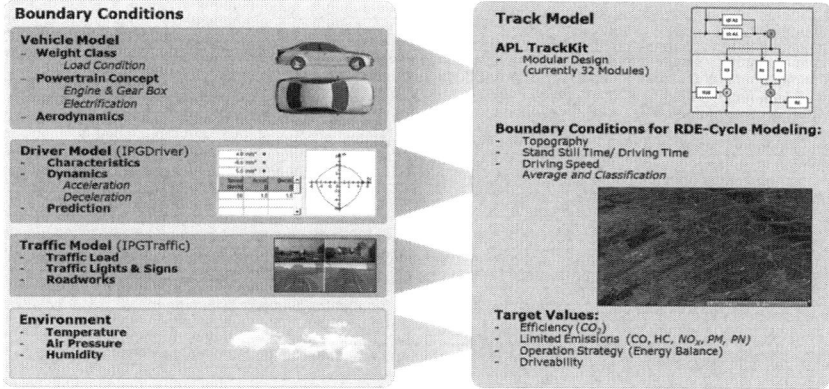

Figure 4: APL TrackKit and simulation environment

For the assessment and analysis of the emissions in the RDE cycle, the occurrence frequency of the individual operating points of the load profile is plotted two-dimensionally over the engine map (Figure 5). In parallel, emission values from the load profile associated with the individual operating points are cumulated and also displayed via the engine map. The emission-relevant dynamic behaviour of the powertrain cannot yet be assessed in this way. Therefore, representative and emission-relevant gradient groups are determined by decomposing the entire load profile into load and speed gradients by evaluating the local minima and maxima. The gradient groups resulting therefrom are multiplied by the frequency of the occurrence and plotted as the emission potential in a map over the speed, load and torque gradient. For further selection, for example, the highest 30% of the emission potential is filtered. This allows representative, emission-critical gradients to be determined and converted into synthetic test cycles.

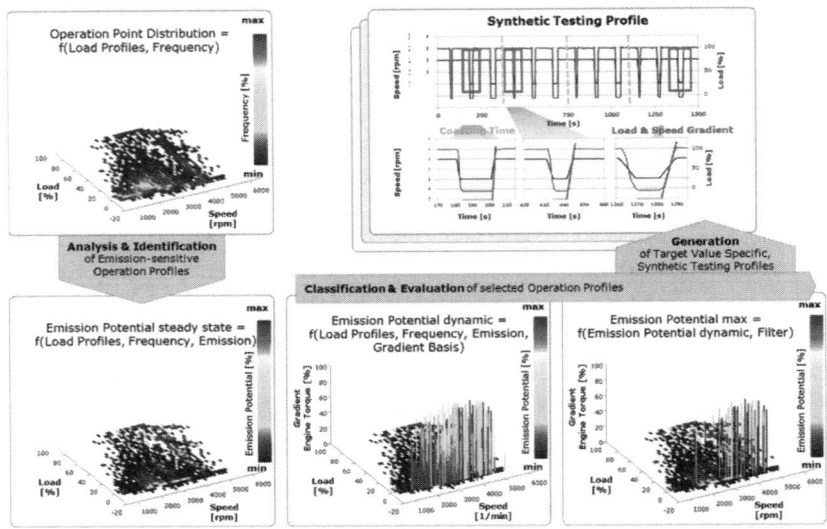

Figure 5: Determination of representative load profiles with high emission potential

3 Exemplary impact analysis on emissions with focus on particle raw emission behaviour

Particle raw emissions are the product of complex internal engine interactions. The mixture formation, in particular the spray breakup of the injected fuel, the injection timing as well as the charge motion is a main influencing factor to avoid an interaction between fuel and component parts.

The operation strategy, the operating fluids and the design of the hardware are very important. In addition to the mentioned influences, aging and wear effects of hardware components and operating fluids also have to be considered (Figure 6). The injection strategy, as an emission relevant part of the entire operating strategy, is essential and has to be defined specifically for every operating point.

The application of online DoE on the test bench allows a reduction in the number of particles by using optimal selected injection patterns (number and position). Due to injection splitting, up to five injection pulses, the spray penetration into the combustion chamber can be reduced, which minimizes the interaction between fuel and close-by parts. In combination with a targeted optimization of the injection pressure, leading to an improvement of the spray breakup and the mixture formation, a simultaneous reduction of rich areas in the combustion chamber can be achieved. In addition to different operating temperatures of the engine, an analysis of the influence between stationary and transient operation has to be done.

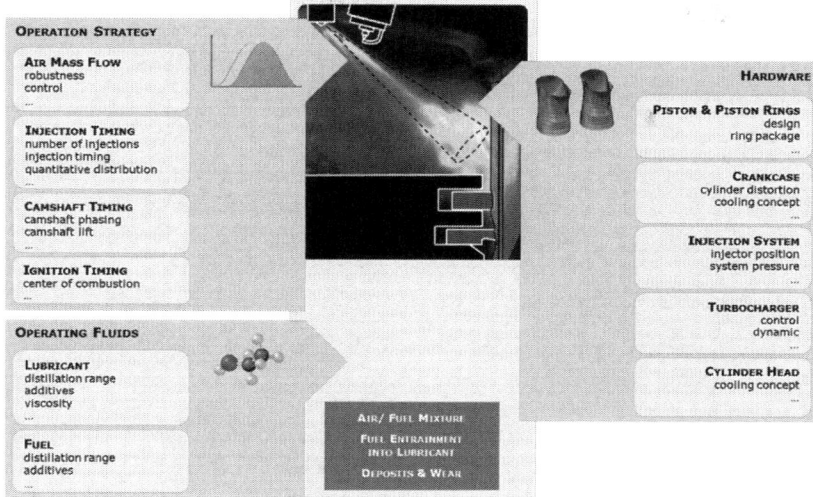

Figure 6: Influencing factors on particle raw emission behaviour

Figure 7 shows a validation of the described stationary phenomena in a high dynamic, representative RDE sequence. The measurements with fuel 2, due to its improved evaporation behaviour in the mid-volatile range, show an advantage of about 30% in terms of particle emission especially in highly transient acceleration phases (detail I & II).

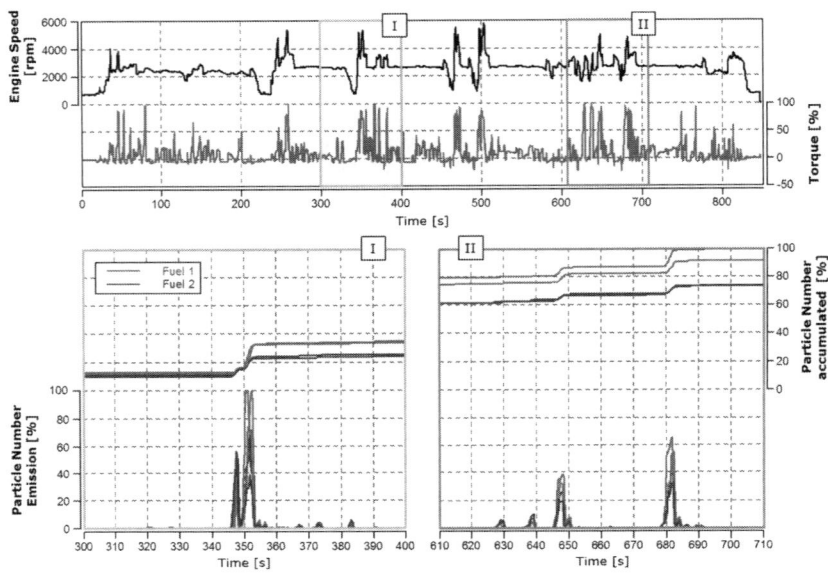

Figure 7: RDE cycle sequence with fuel 1 and fuel 2

Deriving from stationary and transient investigations it can be noted, that even minor changes in fuel properties can lead to significant differences in raw emission behaviour. Considering the fact, that both investigated fuels are within the EU6 certification standards, the challenge of a robust emission behaviour using standard fuels (DIN EN228) is even greater. In addition to a targeted selection of the lubricant, the further focus clearly is a robust calibration with the target values CO_2, particle emission and combustion stability.

4 Efficiency and emission optimization with hybridisation

On the basis of a virtual vehicle model in combination with an empirically determined, dynamic particle emission map, further optimizations are carried out in the simulation environment and validated on the test bench. The intelligent use of electrical energy for boosting using hybrid configurations can contribute to further efficiency and emission optimization. In addition to CO_2 emissions, the focus here is on the sustainable reduction of the particle emission peaks, which mainly are emitted during transient operation.

The greatest potentials for reducing particulate and CO_2 emissions can occur in different operating areas, hence different boosting approaches are necessary. Depending on the fuel quality, for example, additional possibilities could be used here.

With the use of 5 kWh of electrical energy and taking into account a maximum output of the electric motor of 70 kW, different boosting scenarios in a highly dynamic RDE cycle are plotted in Figure 8. The reference is the conventional powertrain without hybridisation. In the development of an operation strategy, a variant can be analysed, taking into account the energy input as a compromise between particle-optimal or CO_2-optimal boost strategies. For example, by a slight shift to the detriment of the CO_2 reduction potential, with the available electrical boost energy a further significant reduction in particle emissions can be achieved.

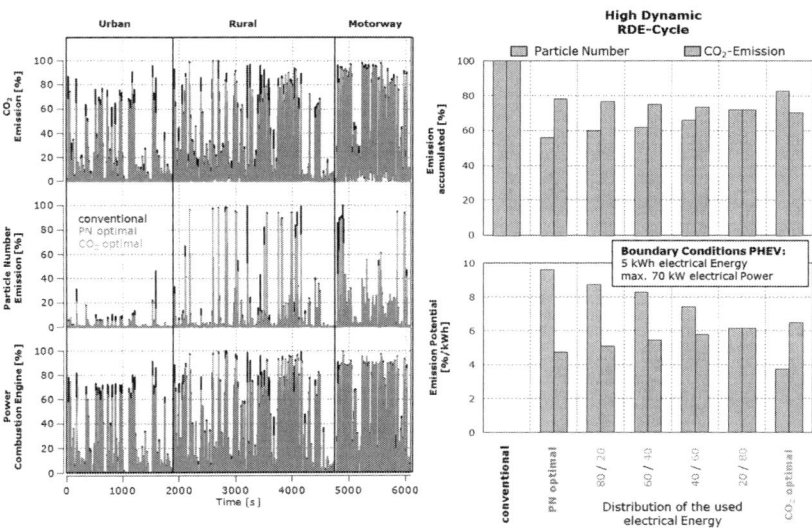

Figure 8: Potential by hybridisation

5 Summary

On the basis of development methods and tools, the APL Group has developed a complex methodology with the focus on RDE compliance over vehicle life. The full potential can best be reached if an understanding of the entire system is already established in an early phase of the development process in order to enable a targeted implementation of individual measures on components in the context of powertrain and vehicle. The described procedure is suitable as a versatile tool for the development, calibration and evaluation of future vehicles, powertrains and exhaust gas aftertreatment systems. The basis is the reproducible transfer of representative real driving cycles to the engine and powertrain test benches. This allows the analysis of the operating parameters and the derivation of functional relationships.

Due to higher dynamic ratios under real driving conditions, new areas for optimization arise in particular in transient operation. In this context, partial electrification creates new degrees of freedom for reducing consumption and emissions.

Literature

[1]   Lensch-Franzen, C.; Gohl, M.; Mink, T.; Schäfer, M.: Impact Analysis of Fuels, Operating Fluids and Combustion Parameters Focus Raw Emission Behaviour. In: MTZ 78 (2017), No. 7-8, p. 62-67

[2]   Hadler, J.; Lensch-Franzen, C.; Gohl, M.; Mink, T.: Emission Reduction A Solution of Lubricant Composition, Calibration and Mechanical Development. In: MTZ 76 (2015), No. 9, p. 30-33

[3]   Hadler, J.; Lensch-Franzen, C.; Gohl, M.; Mink, T.: Concept for Analysing and Optimising Oil Emission. In: MTZ 75 (2014), No. 1, p. 24-28

[4]   Lensch-Franzen, C.; Friedmann, M.; Donn, C; Rohrpasser, C.: Testing with Virtual Prototype Vehicles on the Test Bench. In: ATZ 119 (2017), No. 10 p. 36- 41

Innovative thermal management and waste heat recovery – a combination of technologies for sustainable powertrains

Thomas Arnold, Volker Ambrosius, Matthias Krause

© Springer Fachmedien Wiesbaden GmbH, ein Teil von Springer Nature 2018
J. Liebl (Hrsg.), *Der Antrieb von morgen 2018*, Proceedings,
https://doi.org/10.1007/978-3-658-21419-7_9

Introduction

Driven by the goal to reduce CO_2 pollutions in the European Union by 40% until 2030[1] the fleet average CO_2 emissions of vehicles sold in the EU have to be reduced to 95 g/km in 2021 [2]. A big opportunity to achieve these goals is an increased degree of electrification of powertrains and vehicles as currently under development. So far pure electrically driven vehicles have not reached a level of acceptance that leads so significantly reduced CO_2 emissions. Moreover, current trends as the rapidly growing SUV market share cause increasing CO_2 emissions instead of further reductions.

To sustainably reduce the fuel consumption and emissions in homologation cycles as well as in real world driving conditions the efficiency of internal combustion engines as part of future hybrid powertrains has to be improved. A noticeable reduction in exhaust emissions can be achieved by cutting combustion engines thermal losses.

Measurements at a current, turbocharged gasoline DI engine provided the basis for determining these losses in the WLTC homologation cycle in a C-segment vehicle as shown in figure 1.

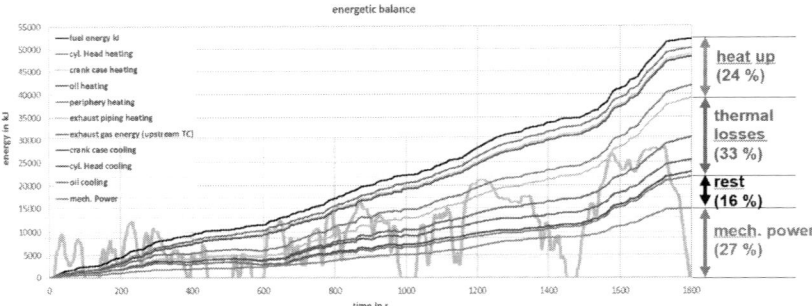

Figure 1: Losses in WLTC [3]

Phase Change Cooling as Measure for reduced Thermal Losses

A promising approach to reduce the thermal losses is to keep the combustion chamber's wall temperature as high as possible during part-load operation

To influence thermal transmittance, it is beneficial to provide the capability of adjusting the combustion chamber's wall temperature independently from the load point over a wider range than is possible with existing cooling processes involving forced convection. These are limited by the maximum permissible fluid temperature of approx. 130°C as well as the maximum heat transfer coefficient α_{wall_limit} primarily dominated by fluid velocity.

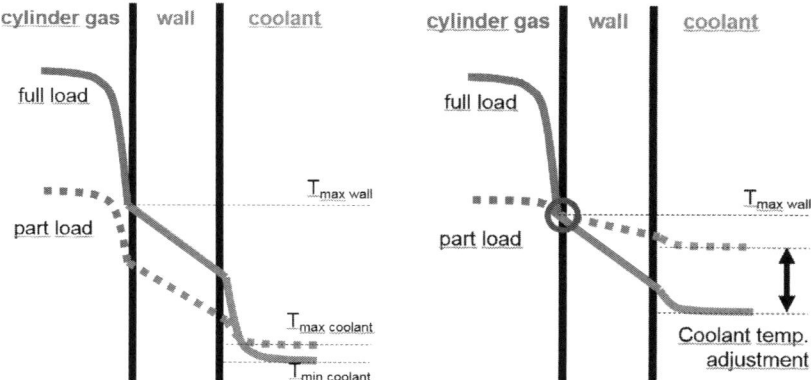

Figure 2: heat conduction and transfer

In designing the convection cooling system for the maximum amount of heat dissipation, adversely low wall temperatures are produced at part load as well as while the engine is warming up. Today, these are improved using measures for flow control. A controlled coolant phase change in the engine, in contrast, promises a significantly wider range for actively adjusting the wall temperature, see figure 3.

Bauteiltemperatur Zylinderkopf

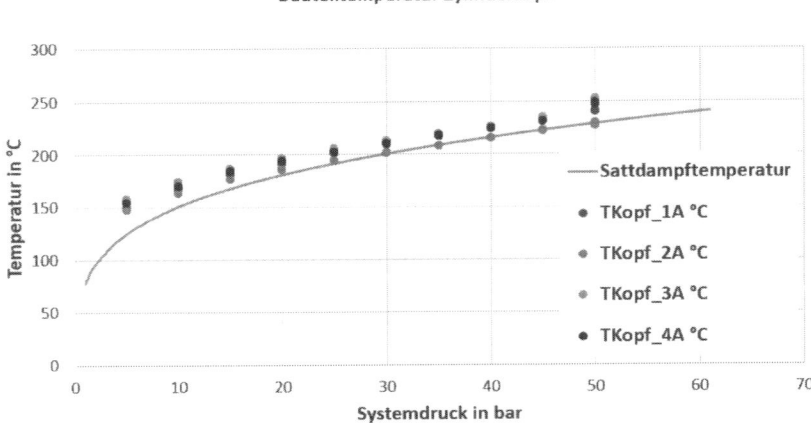

Figure 3: Adjustable wall temperatures [3]

As a result of the far higher wall heat transfer coefficient during the isothermal phase change, very low temperature differences are achieved between coolant and wall, leading to a precisely adjustable and a highly homogeneous wall temperature. In addition to this, very high cooling capacities can be provided in a minimum of package space. [3]

As an result of the increased wall temperatures in part load operation as shown in figure 4, the BSFC can be reduced due to reduced wall heat losses, thermal dethrotteling, reduced coolant pump power and improved EGR capatibility.

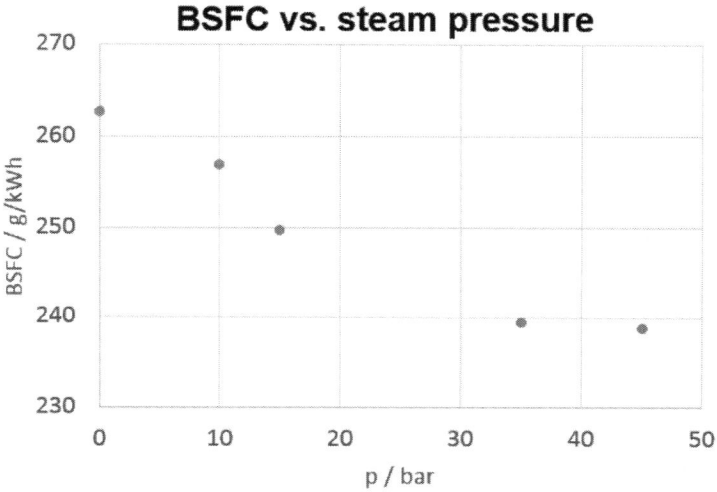

Figure 4: BSFC vs. system pressure @ 1500 rpm / 75 Nm

Combined engine coolant heat plus exhaust heat recovery system

To prove the suitability of IAV's advanced cooling approach the reference engine was equipped with a new designed cylinder head. The base engine for the investigations is characterised in table 1.

Table 1. Reference gasoline ICE for investigations

attribute	value
Engine type	Inline 4 cylinder; Gasoline; turbocharged; direct injection
Max. power	105 kW
Max. torque	230 Nm
displacement	1400 cm³
Emission standard	EU6

For cost reason only the cylinder head was replaced. The cylinder head is the most important part of the engine regarding cooling because of its influence on knocking and engine performance as well as it contains crucial mechanical parts that need cooling to avoid damage or wear. Based on the calculation of achievable steam pressure, steam temperature and the heat flux of all relevant heat sources the system schematic was developed to ensure a safe engine operation and best utilisation of all relevant heat sources.

The executed system schematic is shown in Figure 5.

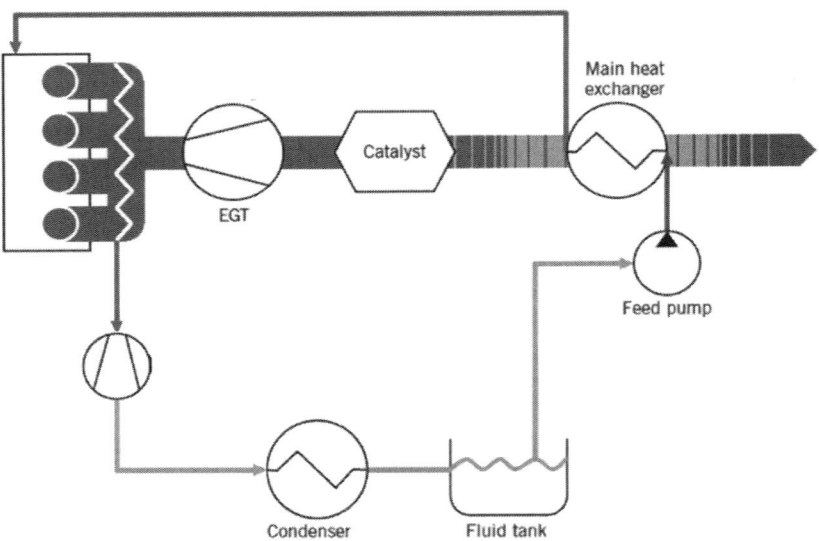

Fig. 5. System schematic for combined waste heat recovery system [3]

For the engine test campaign IAV's passenger car exhaust heat recovery system includ-ing the internally developed pump, tailpipe boiler and piston expander [4] was com-bined with the new designed advanced cooling cylinder head featuring basic thermal measurement to quantify the impact of evaporative cooling on the engine operation as well as its recuperation potential.

The cylinder head was designed in order to keep all functional surfaces and outer ge-ometry unchanged to ensure a possible installation in a demonstrator vehicle.

The final setup on the engine test stand is shown in Figure 6.

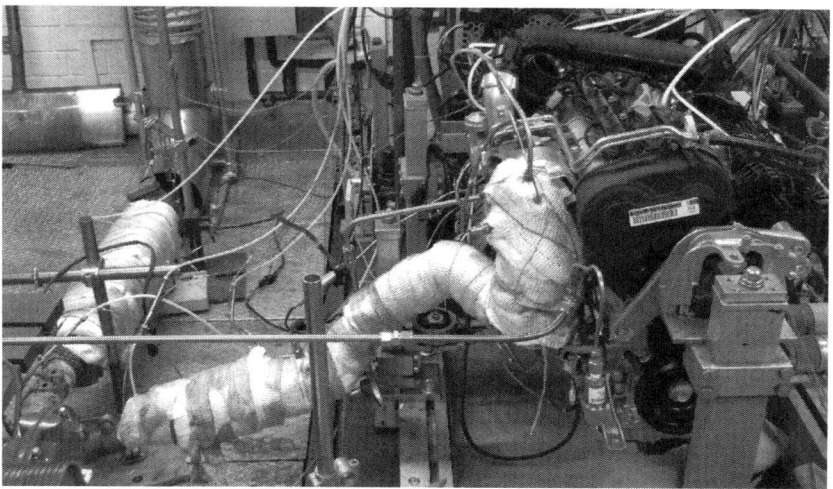

Fig. 6. Engine test stand with combined waste heat recovery system

After the successful commissioning of the combined WHR system measurements in the engine operation map were carried out. The measured points are shown in Figure 7.

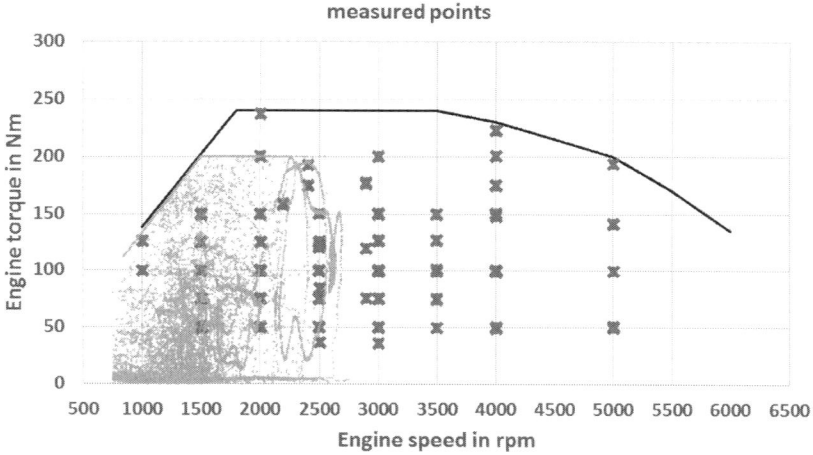

Fig. 7. Measured operation points in engine map

The red crosses represent steady state measurements. The green points represent the operation points of the dynamic WLTC cycle for a given c-segment vehicle that has been measured to quantify the dynamic potential including heating up and controls.

During the steady state measurements several steam parameters such as pressure, degree of superheating and others were varied to quantify their influence on the engine operation (knocking, throttling, BSFC, component temperatures, wall heat losses, power balance shift, ...). The steady state measurements were also compared to the preceding calculations for the simulation model evaluation.

It was seen that the engine equipped with evaporative cooled cylinder head had identical ignition timing and camshaft position compared to the base measurement in the whole engine map.

The measured steam power from exhaust gas and cylinder head is calculated as shown in Eq. (1)

$$P_{Steam} = \dot{m} \times \left(h_{downstream\ cyl.\ head} - h_{AT\ddot{U};20°C} \right) \tag{1}$$

The mass flow is measured downstream the high pressure pump. The enthalpy downstream the cylinder head is calculated using "refprob®". The input for the enthalpy calculation is the measured pressure and temperature at the cylinder head outlet. The reference enthalpy is defined at atmospheric pressure and 20°C.

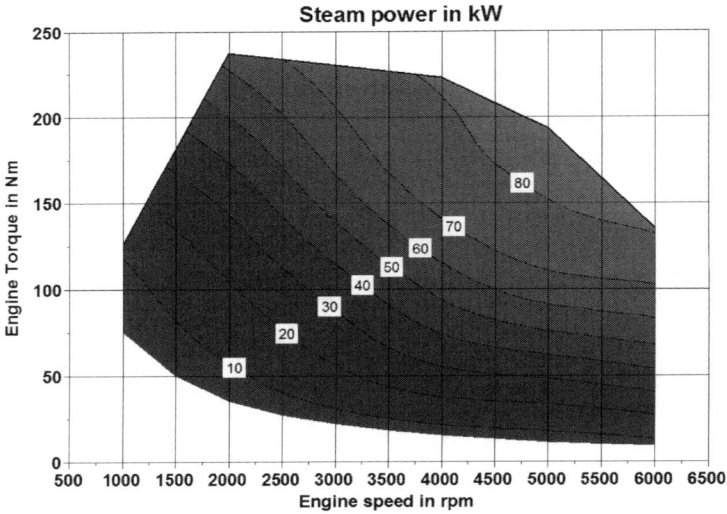

Fig. 8. Measured steam power in engine operation map

Compared to the pure EHR system [4] the increase in generated steam power is (depending on the operating point) from + 55 % up to + 70 %. Due to the increased steam power the mechanical power of the piston expander increases.

Assuming the utilisation of the separately measured crank case cooling power into the combined WHR system leads to a further improvement of BSFC. The specific fuel consumption of the test engine equipped with a complete WHR system including recuperation from exhaust gas, cylinder head and crank case is shown in Figure 9.

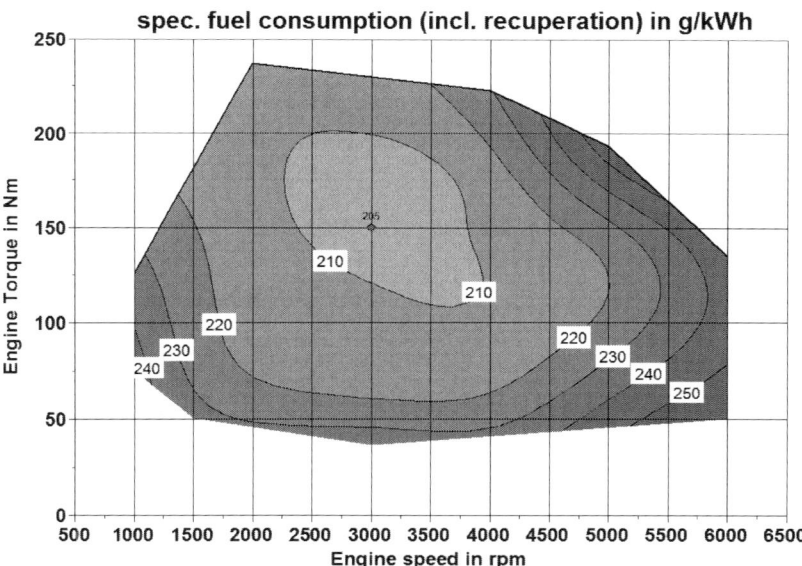

Fig. 9. BSFC with crank case heat included

The inclusion of the crank case cooling into the System leads to BSFC benefits of 12% to 20% depending on the engine operating point. At the reference point (1500 rpm / 75 Nm) the BSFC reduction is determined with 14.6 %.

Based on a validated longitudinal simulation model, the steady state and WLTC measurements of the test engine the fuel consumption potential of the combined waste heat recovery and phase change cooling system was simulated for a C-Segment vehicle as shown in figure 10.

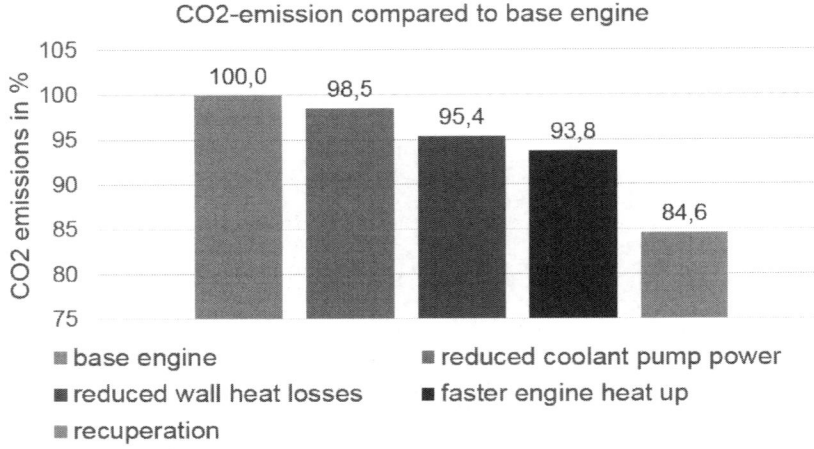

Fig. 10: WLTC fuel consumption benefit

The influence upon the fuel consumption by the reduced water pump power, the reduced wall heat losses and the faster engine heat up by phase change cooling can be seen in figure 11. The overall advantage of the phase change cooling system is 6.2 %.

A system supplemented by an exhaust gas heat exchanger and a piston expander for waste heat recuperation shows a fuel consumption reduction of approximately 15.4 %.

Conclusions

Driven by the necessity of improving the efficiency of future powertrains IAV developed a new approach of combined engine coolant heat and exhaust heat recovery system.

IAV identified the potential of a combined system of exhaust gas and coolant heat recuperation by simulation. Based on the simulations the design of a gasoline engine cylinder head was adapted to fulfil the required functions such as component cooling and recuperation of coolant heat without negative influence on the engine operation.

After successful commissioning of the combined WHR system the engine tests were carried out to prove the system functionality and recuperation potential.

It was shown that utilizing the cylinder head coolant heat only leads to a significantly increased waste heat recovery potential by factor 1.5 to 1.7. Furthermore it was shown that the integration of crank case heat could lead to steady state BSFC benefits of 12 % to 20 %.

It was also shown that evaporative cooling is a suitable technology for cooling turbo charged engines with high specific powers of up to 75 kW/Liter from part load to rated power without restrictions.

References

[1]   https://ec.europa.eu/clima/policies/strategies/2050_de
[2]   http://www.bmub.bund.de/fileadmin/bmu-import/files/pdfs/allgemein/
       application/ pdf/ eu_verordnung_co2_emissionen_pkw.pdf
[3]   M. Weise, T. Arnold, V. Ambrosius, Dr. H. Neukirchner, "Innovatives
       Thermomanagement", MTZ 10/2017
[4]   I. Friedrich, T. Arnold, O. Dingel and H. Neukirchner, „Experimenteller
       Nachweis der Reduktion des Kraftstoffverbrauchs durch IAV WHR-
       Technologie im Pkw", VDI Fachtagung Innovative Antriebe, Dresden,
       23.11.2016

Tagungsbericht

Andreas Burkert

J. Liebl (Hrsg.), *Der Antrieb von morgen 2018*, Proceedings,
https://doi.org/10.1007/978-3-658-21419-7_10

12. Internationale MTZ-Fachtagung
Der Antrieb von morgen

Die Elektromobilität gewährt dem Diesel keinen Aufschub

Fahrzeuge von morgen werden noch lange Zeit verschiedene Antriebsarten nutzen. Auch der Dieselmotor findet seinen Platz, allerdings gewährt ihm die Elektromobilität keinen Aufschub. Mit jedem Kilometer mehr elektrische Reichweite wächst die Kritik am Verbrennungsmotor, wie die emotionale Diskussion auf der Fachkonferenz von ATZlive „Der Antrieb von morgen" zeigt. Dass auch die Brennstoffzelle und synthetische Kraftstoffe auf den Prüfstand gehören, ist der nicht optimalen Energieeffizienz zu verdanken. Dabei wurden in Frankfurt beeindruckende Fortschritte präsentiert.

ELEKTRIFIZIERTER ANTRIEBSSTRANG

Die Ansichten des Vertreters des Hessischen Umweltministeriums Dr. Christian Hey sind der Anlass für eine hitzige Debatte rund um den Antrieb von morgen. Hey ist Abteilungsleiter im Ministerium für Umwelt, Klimaschutz, Landwirtschaft und Verbraucherschutz. Und weil sie als oberste Landesbehörde die Verantwortung für den nachhaltigen Schutz des Menschen und seiner natürlichen Lebensgrundlagen trägt, kann Hey manche Argumente der Automobilhersteller nicht akzeptieren. Beispielsweise verneint er vehement, dass der Dieselmotor Teil einer „Lösung des Klimaproblems" ist. Dies bezeichnet er gar als Falschaussage und nimmt damit die Konfrontation mit Professor Thomas Koch gern in Kauf.

Koch ist seit 2013 Leiter des Instituts für Kolbenmaschinen am Karlsruher Institut für Technologie (KIT) und damit wie kein anderer geeignet, die Faktenlage rund um den Dieselantrieb wissenschaftlich aufzubereiten. Er ist auch enger Berater des Bundesministeriums für Wirtschaft, des Bundesministeriums für Verkehr und des Bundesministeriums für Forschung. Darüber hinaus berät er das Direktorat der EU für RDE (Real Driving Emissions). An diesem Tag aber ist er Gast der internationalen Fachkonferenz „Antrieb von morgen", die von ATZlive organisiert und von Johannes Liebl, Herausgeber der ATZ, und Alexander Heintzel, Chefredakteur der ATZ, inhaltlich geprägt wurde.

DAS NECKARTOR IN STUTTGART ENTSCHEIDET ÜBER DIE ZUKUNFT DES DIESEL

Im Fokus der zweitägigen Konferenz, an der rund 180 Führungskräfte aus der Automobilbranche teilnahmen, stand die Elektrifizierung des Antriebsstrangs. Wohlwissend der enormen Bedeutung für den Standort Deutschland, um die von der Gesetzgebung vorgegebenen Grenzwerte einhalten zu können. Vor allem beim Feinstaub und beim Stickoxid NO_x muss der Verbrennungsmotor sich vorwerfen lassen, das Klima zu „vergiften". Als gerichtsfester Beweis werden dabei auch die Ergebnisse von Deutschlands berühmtester Messstation am Neckartor in Stuttgart vorgelegt. Auch Koch präsentiert in seinem Vortrag die gemessenen zahlreichen „kritischen Überschreitungstage". Schon, um auf die Probleme einer ungenügenden Abgasaufbereitung hinzuweisen, die aber „durch die Euro-6-Norm" seiner Ansicht nach nicht mehr auftreten werden. Vor allem aber will er den Verbrennungsmotor aus der Schusslinie holen. Er ist überzeugt, dass der Beitrag am Feinstaub durch den Straßenverkehr generell viel dominanter ist. Etwa durch „Aufwirbelungen, durch Abrieb, ganz besonders durch Bremsstaub, Kupplungsstaub und derartige Dinge".

„ES GIBT KEIN EMISSIONSPROBLEM BEIM DIESEL"

Und noch eine Erkenntnis kann Koch verkünden: „Wir sehen eine kontinuierliche Verbesserung der Messergebnisse". Damit zeigen Koch zufolge die Software-Updates bereits ebenso Wirkung wie der höhere Anteil von Euro-6-Fahrzeugen, die unter anderem mit einem motornahen SCR-Katalysator arbeiten. Für den KIT-Wissenschaftler ist damit die Sache „vollumfänglich gelöst!". Nun gilt einzig die Herausforderung, „wie wir die Altflotte in irgendeiner Art und Weise verbessern können", so Koch. Er weist darauf hin, dass „keine Technologie so zur Verbesserung der Luftqualität beigetragen hat wie Forschung und Entwicklung am Verbrennungsmotor". Für ihn ist der „Dieselmotor viel, viel besser als sein Ruf", versucht er das Auditorium zu überzeugen und konstatiert, dass es „kein emissionsseitiges Argument gegen den Diesel gibt. Basta!". Eine Steilvorlage für Hey.
Natürlich weiß auch der Vertreter der Landesregierung, dass es relative Effizienzgewinne durch Dieselfahrzeuge gibt. Auch deshalb fordert er von der Bundesregierung die Blaue Plakette, um ein Druckmittel für saubere Diesel zu haben. An dieser Stelle weist Hey aber darauf hin, dass nicht alle von Koch erwähnten Euro-6-Diesel dieses Prädikat erfüllen, sondern derzeit nur jene vier Modelle, die Euro 6d temp erreichen. Und noch etwas stellt er infrage. „Wenn wir allein die gesamte Dieselfahrzeugflotte mit der ottomotorischen vergleichen, stellen wir fest: Da ist kein großer Vorteil. Dieselfahrzeuge sind schwerer und höher motorisiert". Dieser Rebound-Effekt fresse die relativen Emissionsvorteile wieder auf.

MIT HOCHDRUCK RAUS AUS FOSSILEN ENERGIETRÄGERN

Deshalb sei der Dieselmotor nicht die Lösung des Klimaproblems, wie Koch meint. „Das ist eine Falschaussage!", entgegnet Hey und legt nach: „Bis 2050 müssen wir aus den fossilen Energieträgern insgesamt aussteigen, um Klimaneutralität zu erreichen". Damit gelte jede Antriebstechnik, die auf fossilen Energieträgern aufbaut, mittelfristig als überholt. Hilft also nur noch die „Flucht in die Elektromobilität", wie es

Heintzel formuliert. „Sicher nicht", sagt Rüdiger Steiner von Daimler. „Der Verbrennungsmotor wird uns noch über viele, viele Jahre begleiten", so Steiner, der bei dem Stuttgarter Automobilhersteller die Abteilung Simulation und Analyse Motor, AGN, E-Drive Systeme leitet.

DIE GRÖSSTE NATION DER WELT FORDERT EINE EMISSIONSFREIE MOBILITÄT

Als Mitarbeiter eines international tätigen Zulieferers hat Wagner nicht nur die Probleme am Neckartor im Blick. Er verweist auf „Mega Citys, die wir uns in der Größenordnung hier in Deutschland kaum vorstellen können" und wie sie beispielsweise in der Volksrepublik China zu finden sind. Auch deshalb werde dort mit enormer Geschwindigkeit die emissionsfreie Mobilität vorangetrieben, die damit auch die heimische Automobilindustrie unter extrem hohen Druck setze. Den spürt auch Norbert Alt vom Entwicklungsdienstleister FEV.

Dennoch hält der Vorsitzende der Geschäftsführung der FEV Europe auch am Verbrennungsmotor fest. Wohlwissend, dass in naher Zukunft selbst bei positiven Prognosen der Anteil der Elektromobilität etwa nur ein Drittel am weltweiten Fahrzeugbestand betragen wird. Die Elektromobilität aber sei für FEV ein wesentliches Element künftiger Antriebe. Auch weil man „mit einem batterieelektrischen Fahrzeug vom Wirkungsgrad extrem gut unterwegs ist, wenn wir den Strom direkt in mechanische Energie wandeln", so Alt. Und er fügt hinzu, dass der Anteil regenerativer Energie, der derzeit bei etwa 38 % liegt, noch steigen müsse. Für Koch das Stichwort, erneut in die Diskussion einzusteigen.

NADELÖHR IST DIE BATTERIEFERTIGUNG

Für ihn muss die Verkehrswende einhergehen mit einer weitreichenden Energiewende. Das hält Koch für das einzig Sinnvolle, auch um über alle Betrachtungsstufen hinweg eine positive CO2-Bilanz zu erreichen. „Von der reinen CO2-Reduzierungsstrategie wäre das der maximale Ansatz". Dem stimmt auch Hey zu, der nichtsdestotrotz bemängelt, dass andere Länder wesentlich höhere Anteile an Elektrofahrzeugen vorwiesen. Für ihn ist es höchste Zeit „jetzt zu handeln" und damit weg vom Diesel zu gehen.

Damit kann sich Gerald Killmann, der für Toyota in Europa für die Forschung und Entwicklung im Bereich Antriebstechnik und Elektronik zuständig ist, arrangieren. Immerhin haben die Japaner in Europa nur noch einen Dieselanteil von unter 20 %, während der Hybridanteil „deutlich über 40 % liegt". Und dennoch sind es noch knapp eine Million Dieselfahrzeuge, die Toyota weltweit pro Jahr produziert, die vorwiegend in Asien für die Land Cruiser-Klasse eingesetzt werden. Allerdings fertigt Toyota auch bereits 1,5 Millionen Hybridfahrzeuge. Seit Jahren sei da aber die Kapazitätsgrenze erreicht, „weil wir einfach nicht mehr Batterien herstellen können", so Killmann.

DER BRENNSTOFFZELLENANTRIEB ALS DRITTER PFAD DER MOBILITÄT

Da kommt der Mirai gerade recht. Die Limousine aus Japan mit Brennstoffzellenantrieb repräsentiert für Killmann die Mobilität von morgen. Dem Fachpublikum hat er in seinem Vortrag nicht nur die Technik, sondern auch den japanischen Weg zur Brennstoffzelle erklärt. Und den Japanern ist es ernst damit, die „Wasserstoff-Gesellschaft weiter voranzutreiben". Toyota-eigene Patente wurden dazu freigegeben. Ob auch Daimler davon profitiert, steht nicht zur Diskussion. Vielmehr weist Steiner darauf hin, dass auch Daimler mit dem GLC F-Cell den „dritten Pfad" der Mobilität beschreitet. Die Brennstoffzelle sei dabei ein „wesentlicher Aspekt des Elektroantriebs", so Steiner und wehrt sich damit gegen die Behauptung, man habe die Brennstoffzellentechnik verschlafen.

Als Beweis führt er an, dass der GLC-F-Cell-Antrieb „deutlich effizienter, kostengünstiger und kompakter" entwickelt wurde. Und auch Alt von der FEV sieht keine großen Versäumnisse: „Gerade in den letzten Jahren haben wir wieder eine deutlich steigende Anzahl von Entwicklungsprojekten und Vorentwicklungsprojekten zum Thema Brennstoffzelle mit Kunden" durchgeführt. Welcher Antrieb aber wird sich künftig durchsetzen? „Der batterieelektrische oder der Brennstoffzellenantrieb", fragt Liebl. Darauf wagt keiner der Anwesenden eine eindeutige Antwort. Bis auf Hey.

EIN NIEDRIGER ENERGIEEFFIZIENZGRAD IST EIN K.O.-KRITERIUM

Seiner Ansicht nach muss dem batterieelektrischen Antrieb die Zukunft gehören. „Wenn wir jetzt auf die Brennstoffzelle gehen, dann haben wir einen wesentlich niedrigeren Energieeffizienzgrad", kanzelt Hey den Vorschlag ab, den Brennstoffzellenantrieb als die Ideallösung zu postulieren. Er argumentiert, dass „in etwa doppelt so viel Energie verwendet werden muss, um ein Fahrzeug mit einer Brennstoffzelle genauso weit fahren zu lassen wie ein Fahrzeug, das direkt elektrisch betrieben wird", und lässt zudem schon gar nicht das Argument einer überlasteten Stromversorgung gelten. „Wenn wir den Verkehr elektrifizieren würden, dann hätten wir eine zusätzliche Nachfrage von Pi mal Daumen 20 % des Strombedarfs. Das ist machbar."
Doch welcher Antrieb erfüllt schon morgen die strengen Emissionsvorgaben und ist in der Großserie verfügbar? Hier bringt Koch den Dieselmotor wieder ins Spiel. Er verweist dabei auf die Chancen, etwa mit synthetischen Kraftstoffen die Klimaschutzziele erreichen zu können. Als kleine Sensation kann dabei der von Volkswagen entwickelte R33 Blue Diesel gewertet werden. Der regenerative, reststoffbasierte Kraftstoff verbrennt in herkömmlichen Motoren stickoxidneutral, und emittiert Well-to-Wheel über 20 % weniger CO_2 als fossiler Kraftstoff. Doch auch dagegen hat Hey etwas: Synthetische Kraftstoffe erforderten einen um den Faktor 6 höheren Energieeinsatz. Seiner Ansicht nach müssen „die Brennstoffzelle und die synthetischen Kraft-

stoffe reserviert werden für Einsatzzwecke, für die die Elektrifizierung nicht realistisch ist".

WAS MEINEN WIR DAZU?

„Die Brennstoffzelle ist nur zum Teil die Antwort auf die Frage nach dem Antrieb von morgen, da batterieelektrische Mobilität eine bessere Energieeffizienz aufweist und genügend Reichweite zumindest verspricht. Dennoch wird sie als Teil des Antriebsmixes gebraucht. Und auch der Verbrennungsmotor wird auf dem Weg zur rein elektrischen Mobilität noch lange eine wichtige Rolle spielen. Auch dank regenerativer, reststoffbasierter Kraftstoffe."

[Quelle: MTZ 79 (2018), Nr. 5, S. 8ff]

Druck:
Canon Deutschland Business Services GmbH
im Auftrag der KNV-Gruppe
Ferdinand-Jühlke-Str. 7
99095 Erfurt